Lecture Notes in Statistics 111

Edited by P. Bickel, P. Diggle, S. Fienberg, K. Krickeberg, I. Olkin, N. Wermuth, S. Zeger

UNIVERSITY
LIBRARIES

Tallahassee

Springer
New York
Berlin
Heidelberg
Barcelona
Budapest
Hong Kong
London
Milan
Paris
Santa Clara
Singapore
Tokyo

Leon Willenborg
Ton de Waal

Statistical Disclosure
Control in Practice

 Springer

Leon Willenborg
Ton de Waal
Department of Statistical Methods
Division Research and Development
Statistics Netherlands
PO Box 4000
2270 JM Voorburg
The Netherlands

*HA
34
W55
1996*

CIP data available.
Printed on acid-free paper.

Camera ready copy provided by the author.
Printed and bound by Braun-Brumfield, Ann Arbor, MI.
Printed in the United States of America.

9 8 7 6 5 4 3 2 1

ISBN 0-387-94722-1 Springer-Verlag New York Berlin Heidelberg   SPIN 10524404

Preface

The principal part of everything is in the beginning.
—Law Maxim

The main incentive to write the present book on Statistical Disclosure Control, SDC, was a request from the TES Institute (TES=Training of European Statisticians) to the first author to organize a course on SDC as part of their course program for 1995. However, only a few books on the subject seemed to exist and none of those we knew about was suitable for our purposes. Therefore we soon realized that we had to write such a book ourselves. A lucky circumstance was the fact that the authors were at that time involved in preparing a handbook on SDC for internal use at Statistics Netherlands (CBS). The initial idea to simply translate these lecture notes from Dutch into English was quickly abandoned. On closer inspection, it turned out that the internal CBS handbook on SDC contains too many idiosyncrasies and particularities of Statistics Netherlands. In order to serve a wider public we had to write a book with a different perspective. Nevertheless the writing of the internal CBS handbook on SDC proved to be very helpful in writing the present book. They provided us with the themes to discuss – although in a different way – and they also suggested a structure for the present book.

The TES-course is primarily intended for official statisticians who are in one way or another concerned with the dissemination of safe data for statistical use; and so is the present book. It is not written for those who are essentially interested in the statistical and mathematical aspects of the subject, nor for those with a strong legal interest. Yet both types of persons may find one or two things that might interest them.

We tried to make the present book useful for those facing the typical problems of a data disseminator, worrying about the protection of the data against disclosure on the one hand and eagerly trying to satisfy the demands of the data users on the other. Despite this general aim the book is very much biased by the views of the authors, resulting from their own experiences and researches at Statistics Netherlands. It is nevertheless our hope that this bias is not considered too much of a drawback. Indeed, may the reader who is dissatisfied with a procedure, method, *etc.* we describe, find at least enough stimuli and clues in the text to produce a superior alternative!

The aim of the present book is to discuss various aspects associated with disseminating personal or business data collected in censuses or surveys or copied from administrative sources. A problem one then faces is to present the data in such a form that they are useful for statistical research on the one hand and to provide sufficient protection for the individuals or businesses to whom the data refer on the other hand. It is assumed that the data are either microdata (containing information on individuals) or tabular data (consisting of aggregate information). As a provider of statistical data one has to make sure that individual entities cannot be recognized (too easily) by an intruder through inspection of released data. For instance, recognition of an individual could lead to disclosure of confidential information about this individual. This obviously is not in the interest of the individual concerned. But it is neither in the interest of the data provider, since it affects his trustworthiness as a privacy protector.

It should be stressed that safety of a data set is not a "black-or-white" concept, but one with various "gray scales". The reason for this is that data, in particular microdata, that are made absolutely safe may not be useful for certain statistical researches. In such cases it is possible to modify the original data only slightly so as to remove the most "endangered" cases as far as disclosure is concerned. Such "lightly protected" data are then put at the disposal of a select group of researchers. Access to the data could, for instance, only be granted under a license which states the "do's and don'ts" concerning the data and their use. In extreme cases data access is only allowed at the premises of the data provider, *e.g.* a statistical office. The present book summarizes experiences with various modes of access to statistical data, and some of the legal and organizational measures accompanying them.

However the major part of the present book is concerned with the disclosure problem itself: how to define it and how to deal with it in practical circumstances. First of all it should be clear how to properly formulate the disclosure problem for various types of data. This is obviously a necessary requisite in order to be able to make a distinction between data considered safe and data considered unsafe for release to a particular group of users. Such criteria are evidently necessary when modifying unsafe data, *i.e.* not satisfying the criteria, in such a way that the resulting ones are safe,*i.e.* satisfying the criteria. Several well-known techniques, such as em global recoding of variables (also known as collapsing of categories or grouping of variables), local suppression of values, or perturbation of values (such as adding noise, rounding, *etc.*), can be used to produce safe data sets. It is natural to require that the modification should destroy as little information as possible in the data.

The first problem, *i.e.* how to find criteria to distinguish safe from unsafe data, is, in a sense, the hardest one of the two. It can only be considered satisfactorily solved for tabular data. Despite some research this problem is still not satisfactorily solved for microdata. Ideally we would like to have a disclosure risk model that yields, for each record in a microdata set, a probability that the individual to which it refers can be re-identified. But for want of such a model, we have to consider alternative approaches. One such approach, taken by Statistics Netherlands, is considered in more detail in this book. The authors are most familiar with this approach, since they were deeply involved in it. Apart from this the approach seems to be interesting enough in its own right to consider it in some detail as it may pave the way for a disclosure control method based on a re- identification probability-per-record model. In a nutshell the key idea of this approach is the requirement that certain, precisely defined, combinations of characteristics appearing in a microdata file should occur frequently enough in the population. What is "frequently enough" should be defined in terms of suitably chosen threshold values.

This type of criterion is fairly easily to apply in practice, once a non-trivial estimation problem has been solved in case the microdata file contains data from a sample and not from the entire population. A tacit assumption we make is that the data we want to protect have not been perturbed as a result of measurement errors. Although this assumption is obviously unrealistic, it is also a convenient one to make. Otherwise one is forced to formulate a model for such errors, which may not be an easy thing to do, and which, of course, is always more work than ignoring their presence. For a more refined approach one cannot allow to neglect measurement errors in the data. It is clear that the presence of measurement errors decreases the risk of re-identification of individuals. Neglecting such errors therefore tends to overestimate the disclosure risks. For simplicity's sake we have chosen to ignore measurement errors in the present book. We do not have particular bad feelings about this "omission" as the appropriate context for considering the effects of such errors on the disclosure risk, *viz.* an identification probability-per-record model, is currently lacking (in our view). A logical consequence of our negligence of measurement errors would be to abstain from any method that introduces such errors into the data (*i.e.* perturbation methods), or rather, to be more precise, in the identifying part of the data. In fact, that is precisely what we do in case of categorical identifying variables. (In case of continuous identifying variables we do apply such methods as adding noise, however.)

An advantage of non-perturbative techniques, such as global recoding and local suppression, is that one does not have to worry about the integrity (consistency) of the data in the modified data set: these techniques do not affect the integrity of the data; they only reduce its information content.

Because the first problem, *i.e.* how to properly assess the disclosure risk for microdata, the second one, *i.e.* how to efficiently to produce a safe microdata set, can neither be solved in a general way. Therefore the production of a safe microdata set is considered in this book on the basis of the type of safety criterion discussed above, *i.e.* based on the frequency of certain combinations of key values. Producing a safe microdata set in a way that induces as little information loss as possible results in certain optimization problems of the integer programming kind. Only the simplest such problems will actually be considered in this book in order to illustrate the kind of optimization problems one has to solve. A more complete and in-depth treatment would be too technical for the present text, and would not add much to the understanding of the basic problems involved. Furthermore a practitioner is not expected to solve such optimization problems himself. Instead such an individual would benefit most by using specialized SDC software, such as ARGUS, developed at Statistics Netherlands (*cf.* [22] and [58]).

The SDC of tabular data as treated in this book is along the lines of the standard approaches in the literature. In this case we also could distinguish two problems: the definition of the disclosure risk on the one hand and the application of disclosure control measures to produce safe tables on the other. The first problem is not solved here through an elaborate and sophisticated model, but through rather simple models of the behavior of intruders. One such model leads to a well-known type of criterion for unsafe cells in a table, namely the dominance rule. After unsafe cells have been identified in a table one can apply similar techniques to get rid of these cells as one would use in the microdata case: recoding of rows, columns, *etc.* (also referred to as table redesign), cell suppression, or perturbation (addition of noise, rounding). In case cell suppression is applied one may also have to consider additional suppressions when marginal totals are available and it is known that the cell values are limited in one way or another (*e.g.* that they are nonnegative). In general terms, the problem in case of tabular data is similar to that of microdata: how to produce a safe data set with a minimum of information loss (suitably quantified). In the tabular case one also has to solve optimization problems of the integer programming type.

The book contains eight chapters. Chapter 1 gives an introduction to the subject of statistical disclosure control. Chapter 2 elaborates this general introduction. As far as SDC for microdata is concerned the CBS point of view is explained, in a slightly more general way. As far as tabular data are concerned the commonly accepted viewpoint is presented. The discussion in this chapter is essentially at a verbal level, without any technicalities involved. In Chapter 3 the SDC policies at various statistical institutes in the world are described. In some cases a rather detailed discussion of the practices are given. The information presented in this chapter is partly based on a survey carried out a few years ago, and partly on more recent information. In Chapters 4 and 5 SDC issues for microdata are discussed. Chapter 4 gives a non-technical introduction which is elaborated in Chapter 5. Similarly topics on the SDC for tabular data are discussed in Chapters 6 and 7: Chapter 6 being the introduction and Chapter 7 containing various technical details. Persons not interested in the technicalities can skip Chapters 5 and 7. The final chapter, Chapter 8, contains various topics that require additional research. The book concludes with a list of references. A user may want to consult this list when trying to dig more deeply into the subject.

Acknowledgments

We would like to express our gratitude to the following individuals for putting texts at our disposal that we could freely adapt to make them fit into the framework of the present book: Angela Dale (University of Manchester), Joris Nobel and Cor Citteur (both of Statistics Netherlands) and Remco De Vries (formerly of Statistics Netherlands). Angela provided us with a text on the SARs (*cf.* Section 3.5), Joris with a report on the WSA (*cf.* Section 3.4). Cor and Remco produced English translations of portions of texts they originally wrote for the internal CBS handbook on SDC mentioned before, which they co- authored with the present authors.

Furthermore we want to express our appreciation to various staff members, past and present, of Statistics Netherlands. Through their efforts, work or encouragements we benefited, directly or indirectly, in writing this book. In particular we would like to single out (in alphabetical order): Mieke Bemelmans-Spork, Jelke Bethlehem, Piet van Dosselaar, Jan van Gilst, Anco Hundepool, Wil de Jong, Wouter Keller, Peter Kooiman, Robert Mokken, Jeroen Pannekoek, Albert Pieters, the late Albert Verbeek and Philip de Wolf. Several students wrote their master's theses on subjects in the area of statistical disclosure control, while working as trainees at the Department of Statistical Methods. They are: Gary Barnes, René van Gelderen, Jacqueline Geurts, Jeffrey Hoogland, Gerrie de Jonge and Mark-André van Kouwen. Their work was of great value in our understanding various aspects of the subject of this book. We also want to thank several individuals outside Statistics Netherlands who provided us with material and insights that contributed to our understanding of the subject: Matteo Fischetti (University of Padova and University of Udine), Cor Hurkens and Jan Karel Lenstra (both of Eindhoven University of Technology), Dale Robertson (Statistics Canada) and Chris Skinner (University of Southampton). Finally, we would like to thank Hans Dwarshuis for his careful reading of an earlier draft of the manuscript.

It goes without saying that the sole responsibility for the views presented on the following pages rests entirely with the authors.

Voorburg
December, 1995

LEON WILLENBORG
TON DE WAAL

Contents

1
Introduction to Statistical Disclosure Control

Every individual has a place to fill in the world,
and is important in some respect,
whether he chooses to be so or not.
—Hawthorne

1.1 Introduction

Statistical offices basically release two kinds of data, namely tabular data
and microdata. Tabular data are the traditional products of statistical offices. These tables contain aggregated data. Microdata sets are released
only since recently. These microdata sets consist of records with information about individual entities, such as persons or business enterprises, and
have generally been collected by means of a survey. In other words, each
record contains the values of a number of variables for an individual entity.
A microdata set is in fact the raw material that is used to construct tables.
In former days they were actually only used for constructing tables which
were subsequently released. Nowadays, the microdata sets themselves are
also released, although usually in a somewhat adapted form. Because the
disclosure risk for microdata is potentially much higher than that for tables,
this is an important reason for the increasing attention that SDC demands.

In Section 1.2 we sketch the background for data releases, the dilemma
one faces as a data releaser, the threats provided by the widespread use of
personal computers, and legislation that has been produced to accompany
data releases in some countries.

In order to make clear what the disclosure risks are when releasing tables
or microdata, we give a preliminary discussion of this subject in Section 1.3.
In fact it is one of the main themes of the present book. Later chapters
elaborate the issues addressed in this section.

1.2   The increasing importance of SDC

The only goal for any statistical institution has always been, and still is, supplying society with adequate statistical information. Fulfilling this goal is tantamount to aiming at as rich and complete data as possible as the resources permit. At the same time, enhancing quality and detail increases the risk that one or more pieces of data are given in a such a way that they may lead to disclosure of individual data. Such a possibility certainly is not within the intentions for which the statistical data are produced. In this context the contrast between the right of society to information and the right of the individual to privacy is at stake. Briefly, the ultimate goal of SDC is disseminating statistical information in such a way that individual information is sufficiently protected against *recognition* of the subjects to which it refers, while at the same time providing society with as much information as possible under this restriction.

Safeguarding the right to privacy of the individual is a fundamental element in the policy of statistical offices. Several countries have one or the other form of a legal framework dealing with this matter. In some instances a law bears upon a particular survey, such as the Census of the Population (United Kingdom). Relevant laws in the United States and the Netherlands have a wider scope. Title 13 of the United States Code covers the release of microdata for most surveys. In the Netherlands a law regulating the collection and release of economic statistics ordains, among other things, that Statistics Netherlands is accountable for maintaining the confidentiality of the individual (enterprise-) data which have been collected under the provision of this law.

Until relatively recently emphasis in the field of SDC has been mainly on economic statistics, as is illustrated for the Netherlands by the restriction of the relevant legislation to economic statistics. However, during the last two or three decades public concern with issues of privacy of individual persons has been growing fast. Many discussions have been held about the threats to privacy due to the existence of numerous files containing individual records. These discussions were often rather heated, especially in the years in which a Census of the Population was planned. Such was the case in several countries. Under the pressure of these controversies the Census had to be postponed (the former Western Germany), or even abolished (the Netherlands).

Another fact which demonstrates the importance of the personal privacy issue for the general public is the existence of pressure groups. Seminars, debates in the media, and the emergence of journals which are devoted to personal privacy, such as the Privacy Journal in the United States, also bear witness of the growing concern of the privacy issue in society. As a result of this very broad engagement governments in various countries took the initiative for a more systematic handling of privacy issues, as far as a legal setting had not been established already; Sweden, for instance, has a long-standing tradition in safeguarding privacy.

For example, in 1990 the Netherlands witnessed a new law which settles the main aspects of managing files with individual data on persons. The initial motivation for this law dated back from the days of the heated discussions on the 1971 Census of the Population. In this law various measures are prescribed to safeguard the privacy of those concerned. In addition it entitles those concerned the right to inspect and, if necessary, to correct the stored information about them. Such rights, however, would seriously hamper the use of microdata for statistical use. In view of this, and because statistical use of these files does not create direct consequences for those concerned, the "privacy-law" is rather mild in the case of microdata for statistical use.

In the United Kingdom there is a similar law, namely the Data Protection Act of 1984. This law gives subject access rights to individuals to discover, challenge and if necessary correct information about themselves which is held on computer files. The Act places obligations on those who record and use any personal data on computers; they must be open about that use and follow sound and proper practices — the Data Protection Principles — set out in a schedule to the Act. The census databases are subject to the registration provisions of the Data Protection Act, even though names and addresses are not on the computer which holds the main census database. This is because of two reasons: firstly, it is possible for the Census Offices to identify an individual to whom the data relate from other information in the Office's possession; and secondly, post-codes are entered and might in some rural areas lead to a theoretical risk of re-identification. However, since the information is held only for the purpose of preparing statistics and carrying out research, it is exempt from the subject access provisions of the Act. The fact that legislation on personal privacy has taken a firm basis — and not only in the Netherlands and the United Kingdom — generates a stimulus for the statistician to be actively involved in this area.

Apart from legal considerations another motive for maintaining confidentiality comes from the moral obligation to which many statistical offices traditionally have committed themselves. This is necessary to obtain the cooperation of those who are selected in the sample of a survey. Such a behavior is in line with the Declaration on Professional Statistics, formulated by the International Statistical Institute in 1985, as the following quotation witnesses:

> *"Statisticians should take appropriate measures to prevent their data from being published or otherwise released in a form that would allow any subject's identity to be disclosed or inferred."*

Finally, a very practical reason exists for the active participation of a statistical office in privacy protection. Should there be a case in the media in which the privacy in a survey were violated, then this would reduce the willingness of the population to participate in any future survey, not only in the particular one which caused the problem.

Because of the above reasons, statistical offices have dealt with SDC in one way or another ever since their foundation. However, the sharp increase of its importance dates back from relatively recently. This is not only a result of the discussions about the Censuses, and about collecting individual data in general, in the seventies and eighties. At least three other main causes can be given.

Firstly, statistical information is collected in many more fields nowadays than, say, twenty or thirty years ago. Moreover, there has been a continuous quest for a higher quality of statistics. The latter aspect includes matters such as accuracy, finer detail, timeliness and the mutual coherence of different statistics. Simultaneously with the improvement of quality the risk that information of individual respondents is disclosed increases.

Secondly, the ever growing availability and presence in society of computers and sophisticated software presents another ground for serious concern about privacy. Many research-workers from academia and government are now in a position to perform detailed analyses of large data files themselves. They are no longer dependent on the standard tables that a statistical office produces as a summary of the information of a data file. They are able to produce their own tabulations. It is evident, however, that making microdata sets available to these researchers increases the risk that individuals represented in data set can be re-identified.

Thirdly, the overall computerization of society with its numerous data files containing information on individuals presses statistical offices to be very careful. People are aware of the inclusion of a great amount of information about themselves in many files, which in addition can be and, sometimes indeed are, linked with one another. In such an environment a statistical office should take great care to prevent this kind of matching to external files when it releases its own data.

1.3   The Problems

1.3.1   Microdata

Microdata sets, *i.e.* sets of records containing information on individual respondents, are relatively new products of statistical offices. So, the SDC problems caused by the release of these sets are quite new as well. We will sketch the basic problems below.

Whenever a statistical institution wants to release a microdata set it starts by deleting the directly identifying information, such as name, address and telephone number, from each record. Because the records in the released microdata set will not contain directly identifying information, there will be no direct way to determine to whom the information in a certain record pertains. Formally, one could say therefore that the privacy of respondents will not be endangered when such a microdata set is released. Unfortunately, in some cases it is possible to deduce to whom the information contained in a record refers. Suppose, for example, that a statistical institution wants to release a microdata set containing information on the place of residence, the occupation and the criminal past of respondents. Suppose furthermore that a record with the following combination of values occurs in the microdata set: *Residence=Amsterdam, Occupation=Mayor* and *Criminal past=Has a criminal record.* Although, the name or address of the respondent under consideration is not released most persons from the Netherlands will know who the respondent is. In particular, they can conclude that this respondent, the mayor of Amsterdam, has a criminal record.

Even when information on lesser known persons is released privacy problems may arise. Even when, for example, relatively few people would actually know who the mayor of a small village X is, the privacy of this person is in jeopardy when a record containing the combination of values *Residence=X, Occupation=Mayor* and *Criminal past=Has a criminal record* would be released. The reason for this is that everybody knows there is only one mayor in X. This implies, for instance, that it will be relatively easy to obtain this person's name.

In general, the privacy of any respondent scoring on a combination of values that occurs rarely in the population may be endangered. For example, the privacy of the only baker in a small village may be endangered when amongst other information, part of which may be confidential, the occupation and residence of this person are contained in a microdata set. Persons who know that there is only one baker in that village may obtain rather confidential information. Persons who do not know that there is only one baker in that village, may suspect this. In this latter case they may decide to trace this baker, in case they come across very confidential information in his record.

So far, we have only examined combinations of values of occupation and residence, but other combinations of values may be used just as well, of course, to determine to whom a certain record refers. Especially, extremely high (or low) values of quantitative variables may cause disclosure problems. Examples are a person with 10 children and a person with an annual income of 2 million guilders.

Identifying those individuals represented in a microdata file that run a risk to be re-identified is one problem. Another one is how one can modify a microdata file such that the resulting file is safe enough for dissemination. Several general methods are known to modify a microdata file:

- *Global recoding*, also known as *collapsing of categories* for categorical variables. It can be applied to continuous as well as categorical variables, but the result is always a categorical variable. When a categorical variable v in a microdata file is globally recoded categories are combined into new categories. Hence, these new categories are less detailed than the original ones. In the corresponding microdata file, the values for v appearing in the respective records will be replaced by the corresponding new codes. Note that all records are concerned in the microdata file, hence the adjective "global". As an example consider the variable *Age* which can take the values $0, 1, 2, 3, \ldots 99$, each of which stands for the age of a person in years (at a given reference date). Suppose *Age* is recoded to *Age-5cl*, which consists of the categories $1, 2, \ldots, 20$, where 1 stands for the age group 0–4 years, 2 for the age-group 5–9 years, *etc.* If for a microdata file in which *Age* appears this variable is globally recoded to *Age-5cl* then each value of *Age* appearing in the file is replaced by the corresponding value for *Age-5cl*. So for each record in which *Age* takes the value 12, *Age-5cl* takes the value 10-14. The same is true for records where *Age* takes the value 14. In fact for all records in which *Age* takes the value 10, 11, 12, 13 or 14, *Age-5cl* will take the value 10-14. Similarly for all other ages. So the effect of this operation is that less detailed information related to the age of persons becomes available when *Age* is discarded and *Age-5cl* is retained in the microdata file.

- *Local suppression.* This means that the values of a variable v in one or more records are replaced by a *missing* value. The action is called "local" suppression because it does generally not concern all records in a file but only relatively few. Local suppression can be applied to all kinds of variables, continuous as well as categorical. Note that replacing a value of a variable v in a record by a missing variable, has, for that record, the same effect as recoding all categories of v into a single category. For that particular record the action also has the same consequence as deleting the variable from the file.

- *Substitution* (or *perturbation*). This is in fact a collective name for various kinds of modifications which replace (non-missing) values of a variable by other (non-missing) values, that is, without necessarily affecting the definition of the variable as in global recoding. This kind of modification can, in principle, either be carried out globally or locally, but in practice it is only applied locally. Substitution can be viewed as a combination of two operations, carried out one after the other: first (local) suppression, and then imputation. Examples of substitution operations are: addition of "noise" for continuous variables (in which an error term is added to the observed value) and data swapping (in which values of a variable appearing in different records are interchanged). For the imputation step one can use statistical models to predict the value to be imputed from the remaining information in the corresponding record.

Of course, listing the various data modification techniques that can be used does not give a single clue of how to apply them in order to produce a sufficiently safe data set. A substantial part of the present book is about the efficient application of these modifications, given certain goals to be obtained. "Efficient" means that the information loss — suitably quantified — due to the application of the modifications should be minimal. The goal one tries to reach is a microdata set satisfying the rules for safe data. These rules have to be specified by a data disseminator. Part of this book explores the issues that one has to take into account when discussing the "safety" of microdata sets.

In the present book we shall confine our attention essentially to the application of the first two types of data modifications mentioned. The main reason for this is that they are much simpler to apply than substitutions, without a danger of possibly complicated side-effects. The reader should note that there is a fundamental difference between global recoding and local suppression on the one hand and the substitution techniques on the other: global recoding and local suppression ensure the preservation of the integrity of the data. To explain this we introduce the concept of an edit. An edit is a rule that defines, for the variables it involves, which combinations of values are acceptable and which are not. Examples of such edits are: 'if $Age < 16$' then '$Marital\ status = unmarried$' and 'if $Distance\ between\ place$ $of\ residence\ and\ workplace > 0$' then '$Occupation \neq missing$'.

When the values in a record are such that they are acceptable according to a particular edit then we say that this edit is *satisfied* by this record. If an edit is not satisfied by a particular record we say that it is *violated* by this record. A collection of edits can be used to define the integrity of a microdata set: if the values in the records in this data set are such that all edits are satisfied by each record then the integrity of the data is guaranteed (with respect to the given set of edits). The integrity of the original, unprotected, microdata set is usually guaranteed.

In principle, we can distinguish between two kinds of edits: those that do not involve missing values and those that do involve missing values. The first example given above is an edit of the former kind and the second example is an edit of the second kind. Edits of the second kind, *i.e.* those involving missing values, are however hardly applied in practice for microdata sets that are to be released. The reason for this is that a respondent may *refuse to give an answer* to a particular question. For instance, a respondent may refuse to tell his occupation, but may be willing to tell the distance between his place of residence and his workplace. In this case the second edit given above would be violated. Because respondents may refuse to provide answers to certain questions we will consider only edits that do not involve missing values in the remainder of this section.

When none of the edits involve missing values then application of either global recoding or local suppression to an integer microdata set will not affect the integrity. This is not automatically true for substitutions. As an extreme example consider a record with *Occupational status=Unemployed* and *Number of years unemployed=30*. Changing this record such that *Occupational status=Employed* while keeping the number of years unemployed, yields an inconsistency.

Substitutions can yield even more subtle inconsistencies, namely inconsistencies at macro-level. For instance, after substitutions there might be two records with *Occupation=Amsterdam* and *Workplace=Amsterdam* in a microdata set. This is impossible in the Netherlands. The message should be clear: one has to be careful when applying them. It is not that substitutions are considered inadequate for protecting a microdata set, but because they require special attention, we will not deal with them extensively. There is, however, a notable exception to this, and that concerns protecting sampling weights where the addition of noise is recommended as a way to protect these weights (*cf.* Section 4.2.7 and Section 5.5).

More on SDC for microdata sets can be found in Chapters 4 and 5. Because these sets are released only since recently, there is not yet a general agreement on the SDC rules for microdata sets among the releasers. The SDC rules for microdata sets we propose in this book, in particular in Chapter 4, are based on those applied by Statistics Netherlands.

1.3.2   Tables

The best known products of statistical institutions are tables. These tables contain *aggregated* data as the cell values. Because tables contain aggregated data and not data on individual entities, such as persons, households or business enterprises, it may seem that no information about individual entities can be revealed. That this is not necessarily the case will be illustrated using Table 1.1.

TABLE 1.1. Turnover of enterprises × 10 million guilders

	Region A	Region B	Region C	Total
Activity I	11	47	58	116
Activity II	1	15	33	49
Activity III	2	31	20	53
Total	14	93	111	218

At first sight Table 1.1 seems perfectly acceptable for publication. It only contains aggregated data and no data on individual enterprises. However, this conclusion is premature. Suppose for instance that Table 1.1 has been obtained by an integral observation and that there is only one enterprise in region B engaged in activity II. Then Table 1.1 yields that the turnover of this enterprise is therefore equal to 15 × 10 million guilders! So, Table 1.1

cannot be published if we wish to protect the turnover of the only enterprise with activity II in region B. So the lesson from this is that tables with cells that contain the data of only one entity should not be published (provided publication would allow the possibility of making a disclosure about such an entity).

This was easy enough, but there is more to the story. Suppose, for instance, that there is not one enterprise in region B with activity II, but there are two enterprises. In this case, no information on an individual enterprise can be revealed. So, the table appears to be safe. However, in that case each of the enterprises in region B with activity II can disclose the turnover of the other using Table 1.1. Again individual information can be disclosed. Instead of demanding that at least two entities contribute to a cell value one should demand that at least three or more entities contribute to this value. Unfortunately, even this may not be enough.

Suppose now that there are 10 enterprises in region B with activity II, and that one of the enterprises in region B with activity II contributes 95% to the total value of the cell. In that case one can make a reliable estimate of the turnover of this enterprise if one knows that the contribution of this enterprise to the cell is very high. When an entity contributes much to the value of a cell, we say that this entity dominates this cell. So a cell should not be dominated by the contribution of one contributing entity. As before we may extend this and demand that the value of a cell should not be dominated by the total contribution of two entities. This avoids that each of these contributing entities can estimate the contribution of the other one too accurately.

So far we have examined whether or not a certain cell value may be published. Now we examine the possible SDC measures one has at one's disposal to protect a table with cell values that may not be published. The most frequently applied technique to protect such a table is simply by deleting all values of the cells that may not be published. This is called (primary) cell suppression. Suppose that the total turnover of enterprises in region B with activity II is the only cell value that may not be published in Table 1.1. In that case we can delete this value from the table. The result is given by Table 1.2.

TABLE 1.2. Turnover of enterprises × 10 million guilders

	Region A	Region B	Region C	Total
Activity I	11	47	58	116
Activity II	1	×	33	49
Activity III	2	31	20	53
Total	14	93	111	218

A quick glance at Table 1.2 is sufficient to convince oneself that Table 1.2 is not protected at all. The total turnover of the enterprises in region B with activity II can easily be calculated by using the marginal totals. This total turnover is given by 51-3-33=93-47-31 $= 15 \times 10$ millions of guilders. So, when marginal totals are given in a table additional cell values may have to be suppressed to prevent the re-calculation of values that may not be published. In each row and each column at least two cell values should be suppressed. A possible way to prevent the re-calculation the total turnover of the enterprises in region B with activity II is demonstrated in Table 1.3.

TABLE 1.3. Turnover of enterprises $\times$ 10 millions of guilders

	Region A	Region B	Region C	Total
Activity I	11	47	58	116
Activity II	$\times$	$\times$	33	49
Activity III	$\times$	$\times$	20	53
Total	14	93	111	218

In this case the total turnover of the enterprises in region B with activity II cannot be re-calculated. So, it seems that Table 1.3 is sufficiently protected. However, this conclusion is again premature. The turnover of an enterprise is a nonnegative number by definition. Taking the nonnegativity of the cell values of Table 1.3 into consideration enables one to compute the ranges in which the suppressed values must lie. In case of Table 1.3 one can show that the total turnover of enterprises in region B with activity II is at least 130 millions of guilders and at most 160 millions of guilders. By assuming it to be 145 millions of guilders one can be sure that the difference with the actual value is at most about 10%. This difference may be considered too small. When this is the case another table should be published instead of Table 1.3, for instance one like Table 1.4.

TABLE 1.4. Turnover of enterprises $\times$ 10 millions of guilders

	Region A	Region B	Region C	Total
Activity I	11	$\times$	$\times$	116
Activity II	1	$\times$	$\times$	49
Activity III	2	31	20	53
Total	14	93	111	218

In this case one can only deduce that the total turnover of the enterprises in region B with activity II lies between 0 and 480 millions of guilders. Table 1.4 is therefore much better protected against disclosure than Table 1.3.

Instead of suppressing cells in tables one can also consider redesigning tables, *i.e.* combining rows, columns, *etc.* This operation is similar to global recoding in a microdata set. For instance, in the example table considered one could protect the cell corresponding to Region B and activity II by combining Region B and Region C, or activity II and I, whatever produces a safe table. In Table 1.5 the result of combining Region B and Region C into Region B+C is shown.

TABLE 1.5. Turnover of enterprises × 10 millions of guilders

	Region A	Region B+C	Total
Activity I	11	105	116
Activity II	1	48	49
Activity III	2	51	53
Total	14	204	218

Whether the cell corresponding to Region B+C and Activity II is safe should be checked by means of a dominance rule. Note that if all other cells in columns B and C in the original Table 1.1 were safe then the collapsing of columns B and C results only in a loss of information for those other cells. As in the case of microdata recoding is a first step when protecting a table. When there are relatively many unsafe cells in a table then this should be interpreted that, in view of the dominance rule used, the tabular layout is too detailed and should be adjusted. In order to avoid that through using table redesign as the only SDC measure results in an unduly high information loss, one could use cell suppression in a final step. But the number of cells to be suppressed should not be too high.

The reader should be aware of the similarities and differences in the protection of microdata sets and tables. Similar are the recoding and suppression actions. Different is the situation of a table with marginals which give additional information that is lacking in the case of microdata. For a complete table, *i.e.* one without missing cells, these marginals are in fact redundant information. The presence of this additional, redundant information is the major source of difficulties when producing safe tables with a minimum information loss. The example of a 2-dimensional table with marginals is of a type that has been studied traditionally in the literature on the subject. The reader should be aware that this setting is only a special case of a situation in which there is a set of so-called linked tables, all derived from the same base file using different aggregations (*cf.* Section 6.6). The linked tables setting is the general and natural one for discussing the secondary cell suppression problem and the table redesign problem of tables. Applying a dominance rule to these tables yields the sensitive cells. The question then is to protect all these sensitive cells by tabular redesigning and by secondary suppressions in such a way that the value of no primary suppressed cell can be re-calculated with too high a precision.

More on suppression can be found in Chapters 6 and 7. In these chapters also an alternative technique to protect tables, rounding, is examined. The rules for SDC of tabular data can be found in Chapter 6. In contrast to the SDC rules for microdata there is a general agreement on the rules for tabular data among the data releasers.

2
Principles

A good principle not rightly understood
may prove as hurtful as a bad.
—Milton

2.1 Introduction

As we have seen statistical offices may release two types of data: tabular
data and microdata. Microdata consist of records, each containing information about an individual entity such as a person, household, business, *et
cetera.* We assume that survey and census data, after being collected and
"cleaned", finally are collected in a microdata file. Such a file can serve for
the production of tables or it can serve as a basis for microdata files suitable
for release by a statistical office. Both the tabular data and the microdata
that are to be released have to be of such a quality that no individual entity
represented in the data runs too great a risk (in some cases no risk at all)
of being compromised. This means that the risk that some (confidential)
information about such an individual can be deduced or estimated from the
released data with a sufficiently high precision should be kept sufficiently
low.

Controlling this risk is the essence of SDC. The aim is to find the right
balance between disclosure risk on the one hand and richness of information
content of the data (*e.g.* for research purposes) on the other. In practice this
means that one first has to define, for a particular type of data (that are to
be released under particular conditions to a particular group of users) what
the criteria are that such data have to satisfy in order to be considered safe
for release. Given such criteria, the problem then is to modify data that do
not satisfy these criteria in such a way that the resulting data do. Of course,
the challenge is to modify the data in such a way that the information loss

is minimized. When "information loss" is formally defined then the entire data modification enterprise can be formulated as a set of optimization problems. Each problem has its own constraints, in terms of SDC measures to be used to modify the data or in terms of other restrictions imposed by the data disseminator. Formulating and solving these problems is a rather technical matter.

The present chapter is intended as a non-technical introduction to SDC problems for microdata as well as for tabular data. The stress is on relevant concepts and general ideas. In subsequent chapters the general ideas will be elaborated further.

Before we can begin with describing the principles of SDC some basic concepts have to be introduced. This will be done in Section 2.2. Sections 2.3, 2.4 and 2.5 are devoted to microdata. As in the case of tables the role of a user of microdata is twofold: on the one hand he is a client of the statistical office, on the other hand he is a potential attacker. The consequences of this double role are examined in Section 2.3. The basic philosophy of SDC for microdata is discussed in Section 2.4. In particular we explain why combinations of values that occur rarely in the population may lead to disclosure. In Section 2.5 we describe three situations. We start with the ideal situation in which we would possess a realistic and tractable model to calculate a re-identification probability for each record in a file. A somewhat less ideal situation is also described: in this case we only possess a model for a re-identification probability for an entire microdata set, *i.e.* a model for the probability that an arbitrary record of this microdata set is re-identified. We end the section by facing reality: at the moment we do not have a reliable model, yielding a re-identification probability per record. So in order to be able to carry on in practice we cannot be guided by re-identification models and we rely on heuristic arguments. Finally, in Section 2.6, SDC for tables is discussed, in terms of: definition of sensitivity criteria, secondary (cell) suppression and rounding. We only examine the main ideas, leaving the details for Chapters 6 and 7.

2.2   Basic Concepts

In this section a number of basic SDC concepts are defined. We will assume that a statistical office wants to release a table or a microdata set containing records of (a sample of) the population. In the latter case each record contains information about an individual entity. Such an entity could be a person, a household or a business enterprise. In the rest of this chapter we will usually consider the individual entity to be a person, although this is not essential for the discussion.

The two most important concepts in the field of SDC are *re-identification* and *disclosure*. Following Skinner (*cf.* [66]) we make a distinction between two kinds of disclosure. *Re-identification disclosure* occurs if the attacker is able to deduce the value of a sensitive variable for the target individual after this individual has been re-identified. *Prediction disclosure* (or *attribute disclosure*) occurs if the data enable the attacker to predict the value of a sensitive variable for some target individual with some degree of confidence. For prediction disclosure it is not necessary that re-identification has taken place. Most research so far has concentrated on re-identification disclosure. Unless stated otherwise we will use the term disclosure to indicate re-identification disclosure throughout this book. For more information on prediction disclosure we refer to [66], [70], and [32].

Now, let us define what is meant by an identifying variable. A variable is called *identifying* if it can serve, alone or in combination with other variables, to re-identify some respondents by some user of the data. Examples of identifying variables are age, occupation and education. A subset of the set of identifying variables is the set of *formal* (or *direct*) *identifiers*. Examples of formal identifiers are name, address and social security number. Formal identifiers must have been removed from a microdata set prior to its release, for otherwise re-identification is very easy. Other identifiers do not necessarily have to be removed from the microdata set. A combination of identifying variables is called a *key*. The identifying variables that together constitute a key are also called *key variables*. Dalenius (*cf.* [14]) calls the variables in a key *quasi-identifiers*. This is in fact a more appropriate name than *key variables*, because of the specific meaning attached to the latter term in database theory, where it is used to indicate what we call formal identifiers. We shall however stick to the name *key variables* in the present book.

In practice, determining whether or not a variable is identifying is a difficult problem which has been neglected in the literature. Of course, no limitative list of intrinsically identifying variables exists, but neither does an unambiguous and well-defined set of rules to determine such variables. Such rules should also specify whether variables that have been derived from identifying variables are also identifying. Without such rules selecting a set of identifying variables has to be based on intuition. Statistics Netherlands applies some criteria, for instance whether the property a variable refers to is visible or can be known by an outsider. If one wants to avoid that the data to be released will be matched with an existing register, then, of course, one should consider the variables that can potentially be used for this purpose as identifiers.

As a practical solution to this problem of deciding which variables to consider as identifying one could ask a knowledgeable person to decide for each variable in a data file whether it is identifying or not, and accept his decision as binding. In the remainder of this chapter we shall assume that a set of keys has been determined in one way or another.

For tables one usually considers the variables that define the cells of the table to be identifying in practice. Although easy to apply this does not always seem to be appropriate and may lead to the circumstance that a table is sometimes protected too severely. This is particularly the case if some protection procedure is applied mechanically without contemplating the meaning of the information in the table first, in the light of possible disclosure risks.

The counterparts of identifying variables are the *sensitive*, or *confidential*, *variables*. A variable is called sensitive (or confidential) if some of the values represent characteristics a respondent would not like to be revealed. In principle, Statistics Netherlands considers all variables to be sensitive, but in practice some variables are considered to be more sensitive than others. Even more than in the case of identifying variables, specifying which variables are very sensitive and which are less so is a rather vague problem. It is essentially a matter of taste and "public opinion" and is often very dependent on a particular culture. For instance, the variables *Sexual behavior* and *Criminal past* are generally considered sensitive. But, as Keller and Bethlehem (*cf.* [42]) indicate, a variable such as *Income* is considered sensitive in the Netherlands, whereas in Sweden it is not. Moreover, there are variables which should be considered both identifying and sensitive. An example of such a variable is ethnic membership. Income is another example in some countries: it could, for instance, be used to identify a person in a village of whom it is generally known that he has a very high income, without knowing exactly how high it is; once this person is re-identified on the basis of only a rough knowledge of his income one can learn what his exact income is, say to the last cent, penny or whatever.

For tables one often considers the variable of which the value is published in the cells of the table to be sensitive. A practical advantage of this approach is that it is very easy to apply. A disadvantage may be that one protects a table too severely, namely in those cases where the published cell values are not really sensitive.

By using information about the identifying variables an attacker can try to disclose information about sensitive variables. Note that this way of disclosure is only possible in case the link between the identifying variables and the sensitive variables has not been perturbed, *e.g.* by adding noise to the data. That is, if someone has been re-identified by an intruder, this intruder must be sure that the remaining information about the re-identified person is reliable, otherwise he cannot be sure that he has revealed anything. Note that this raises a problem for a data disseminator as well as giving him opportunities to produce "safe" data. The problem is that of a "would be" disclosure of a person whom an attacker seems to have re-identified. In case the re-identification is incorrect it is unlikely that the revealed confidential information about this person is correct. But also in case the re-identification is in itself correct, if the sensitive information of this person was perturbed then no reliable information has been obtained. But unless the intruder is aware of this, there is a really awkward situation. It can be avoided when the data disseminator by clearly states that the data have been perturbed and that *any resemblance to any actual character is accidental and not intended.* The lesson to be learned from this is that the data disseminator can increase the safety of a data set by perturbing either the scores on the key variables or the scores on the sensitive variables (or both). In the next section we shall briefly discuss these perturbation techniques, but mainly as an introduction to non-perturbative ones. These latter techniques will in fact be studied more extensively later on in this book.

As statistical disclosure control for microdata differs from that for tabular data — despite things in common — we discuss both cases separately. We begin by examining the situation for microdata.

2.3   Preliminaries on the SDC for Microdata

The user of microdata plays a very important role in the theory of SDC. His role is twofold. On the one hand he is a customer of the statistical office, on the other hand he is a potential attacker. As a customer of the statistical office, the user should be satisfied with the quality of the data. The user is usually not interested in individual records, but only in statistical results which can be drawn from the total set of records. For instance, he wants to examine tables he has produced himself from the microdata set.

Because a microdata set is meant for statistical analysis it is, in principle, not necessary that each record in the set is correct. The microdata set can be considered as a sample from a possibly very high-dimensional distribution. The important point is that the sample as a whole should give a good impression of the population distribution, not that every sample point is correct, in the sense of in exact agreement with reality. Therefore the statistical office sometimes has the possibility to perturb records by adding noise or by swapping parts of records between different records, in order to reduce the risk of re-identification. How perturbation of the records may help reduce the disclosure risk is discussed in the previous section. If the statistical office also wants to guarantee the statistical quality of, for instance, the tables the user wants to examine, without knowing exactly which ones, then it faces a difficult problem. Meeting both goals in the extreme, *i.e.* safe data and data with exactly the same statistical properties, simultaneously by perturbing the microdata seems impossible, since they are opposite. Only by finding a balance between either goal the problem may be solved. We shall not dwell upon this matter any further. A reader interested in a perturbation technique like data swapping is referred to [15] and [59]. Adding noise is discussed in [37] for instance.

In SDC some of the users of the data should be looked upon as potential attackers. It is useful to consider the ways in which disclosure can take place. An attacker tries to match, consciously or subconsciously, records from the microdata set with records from an identification file or with individuals from his circle of acquaintances. An *identification file* is a microdata

file containing formal identifiers and some other identifiers appearing also
in the microdata set under consideration. The latter identifiers may be used
to match records from the microdata set with records from the identification file. After matching the formal identifiers can be used to determine
whose record has been matched and the sensitive variables can be used to
disclose information about this person. A *circle of acquaintances* is the set
of persons in the population for which the attacker knows the values on
the key variables in the microdata set. So, a circle of acquaintances could
actually be an identification file, and *vice versa* an identification file could
also be a circle of acquaintances. In the rest of this chapter *identification
file* and *circle of acquaintances* will be considered as synonyms.

Because a potential attacker tries to match, consciously or subconsciously, records from the microdata set with records from an identification
file, the theory of re-identification is closely related to the theory of exact
matching (*cf.* [36]). Therefore there is a close link between the theory of
exact matching and the theory that tries to quantify disclosure risks via
re-identification.

For re-identification of a record of a respondent to occur the following
conditions have to be satisfied:

C1. The respondent is unique on the values of the key K.

C2. The respondent belongs to an identification file or a circle of acquaintances of the attacker.

C3. The respondent is an element of the sample.

C4. The attacker knows that the record is unique in the population on
the key K.

C5. The attacker comes across the record in the microdata set.

C6. The attacker recognizes the record of the respondent.

Whenever one of the conditions C1 to C6 does not hold, re-identification
cannot be accomplished with absolute certainty. If condition C1 or C4
do not hold, then a matching can be made but the attacker cannot be
sure that this leads to a correct re-identification. Note that condition C6
implies that even in case there are data incompatibilities, *e.g.* caused by
measurement errors or coding errors, between the released microdata set
and an identification file the attacker succeeds to recognize the record of
the respondent.

It is clear from the conditions C1 to C6 that any good model for the
risk of re-identification should incorporate aspects of both the data set
and the user. When a Dutch microdata set is used by someone in, say,
China who is unfamiliar with the Dutch population, then the risk of re-
identification is zero. In order to re-identify someone in a microdata set

it is necessary to acquire sufficient knowledge about the population. The amount of work needed to acquire such knowledge is proportional to the safety of the microdata set.

2.4   The Philosophy of SDC for Microdata

It seems likely that the attention of a potential attacker is drawn by combinations of scores on identifying variables that are rare in the sample or in the population. Combinations that occur quite often are less likely to trigger his curiosity. If he tries to match records consciously, then he will probably try to do this for key values that occur only a few times. If the matching is done subconsciously and the user knows an acquaintance with a rare key value, then a record with that particular key value may cause him to consider the possibility that this record belongs to this acquaintance. Moreover, the probability of a correct match is higher in case the number of persons that score on the matching key value is smaller. Finally, it is also very likely that among the persons that score on a rare key value there are many uniques if the key is extended with an additional variable. Records that score on such rare combinations of identifying variables are therefore more likely to be re-identified.

A way to visualize the situation is the following. Suppose one wants to examine a particular key K. Consider for each record of the microdata set only the values on the key variables of K. Now, each record can be presented as a point in a high-dimensional space. The dimension of this space is equal to the number of key variables of K. All the records together constitute a cloud of points in this high-dimensional space. Records that are the ones most likely to be re-identified are the outliers or the isolated points of this cloud of points. A possible way to evaluate a re-identification risk per record could be based on establishing which records are the outliers or the isolated points. In order to be able to talk about "outliers" and "isolated points" one should have a metric, which allows one to calculate distances between points. Multivariate techniques such as (multiple) correspondence analysis, or cluster analysis might be employed to find suitable metrics.

If no metric is readily available one could consider the population uniques as the closest thing to outliers. One should be aware that uniqueness is neither sufficient nor necessary for re-identification. If a person is unique in the population, but nobody else is aware of this (cf. C4 in Section 2.3), then this person will not be re-identified by anybody. If on the other hand a person is not unique in the population, but there is only one other person in the population with the same key, then this other person is, in principle, able to re-identify him. Furthermore, if a person is not unique in the population, but belongs to a group of persons with (almost) the same score on a particular sensitive variable, then sensitive information can be disclosed about this individual without actually having been re-identified.

Finally, suppose a respondent is not unique, but belongs to a small group
of people. Suppose furthermore that the attacker happens to have some
information about this person which is not considered to be identifying
by the statistical office, but which is contained in the released microdata
set. Then it is very well possible that the respondent is unique on the key
combined with the new information. Therefore, it is possible that a person is re-identified although he is not unique in the population on the key
variables.

Yet another argument against considering population uniqueness unre-
servedly as a departing point for assessing a disclosure risk is that it does
not discriminate among the uniques. As a variation on a slogan in George
Orwell's Animal Farm we might say that "Some of us are more unique
than others". That is, some people are unique even on a small set of key
variables whereas "everyman" needs a few more in order to make him or
her unique[1]. And it can be argued that the more unique ones are the more
vulnerable ones — in terms of re-identification —than the less unique ones.
For instance, a very unique person is more likely to be re-identified despite
some errors in the recording of his or her information than a less unique
person. For a less unique person a few measurement errors make him look
like somebody else in the population. Another argument in support of the
thesis could be that persons who are "very unique" tend to be better known
in the population. Because the "very unique" persons are the most likely
ones to be re-identified and these persons are unique on a low-dimensional
key already it is sufficient to examine low-dimensional keys only.

A practical problem in dealing with population uniqueness is that it is
hard to verify. Therefore, we relax the condition that we should only consider uniqueness. Instead we propose to consider "rareness" in the population as the important factor defining the re-identification risk. This has the
additional advantage that even when an attacker uses a somewhat higher-
dimensional key in his evil attempts to disclose information than the keys
considered by the data releaser, he is in many cases still not able to find
unique persons in the population on that key.

The "rareness" should refer to properties characterized by only very few
key variables. A probability that information from a particular respondent,
whose data are included in a microdata set, is disclosed should reflect the
"rareness" of the key value of this respondent's record. A probability for
the event that information from an arbitrary respondent is disclosed should
reflect the "overall rareness" of the records in the data set. If there are many
records in a microdata set of which part of the key values are rare, then
the probability of disclosure for this data set should be high.

---

[1] Ultimately, of course, we are all unique.

2.5   The World According to a Releaser of Microdata

In an ideal world (as far as SDC is concerned) a releaser of microdata would be able to determine a probability for each record to express the risk of re-identification. Such a risk per record would enable this releaser to adopt the following strategy. Firstly, he orders the records according to their risk of re-identification with respect to the entire key. Secondly, he selects a maximum risk he is willing to accept. Finally, he modifies all the records for which the risk of re-identification (with respect to the key chosen) is too high. The modifications can be stopped as soon as each record (modified or otherwise) has a disclosure risk below the chosen threshold value.

Unfortunately, we do not live in such an ideal world at the moment. A step towards the ideal situation has been made by Paaß and Wauschkuhn (cf. [52]). In [52] it is assumed that an attacker has both a microdata file released by a statistical office and an identification file at his disposal. Between both files there may be many data incompatibilities. These incompatibilities are caused by measurement errors, coding errors, by the use of different definitions of categories, et cetera. By assuming a probability distribution for these data incompatibilities and a disclosure scenario Paaß and Wauschkuhn develop a sophisticated model to estimate the probability that a specific record from the microdata file can be re-identified.

Paaß and Wauschkuhn distinguish between six different scenarios. Each scenario corresponds to a special kind of attacker, with his own kind of population information and his own motifs for attempting to disclose confidential information. The number of records in the identification file and the information content of this file depend on the chosen scenario. An example of such a scenario is the journalist scenario, where a journalist selects records with extreme attribute combinations in order to re-identify respondents with the aim of showing that the statistical office fails to secure the privacy of its respondents. Another example is the public prosecutor scenario, where the prosecutor tries to disclose a businessman's property data. In case of the journalist scenario it is assumed that the journalist has 10 variables at his disposal to re-identify an unspecified person, in case of the public prosecutor scenario it is assumed that 68 variables are available to the prosecutor to re-identify data of a specific person. Paaß and Wauschkuhn apply their method in order to match records from the identification file with records from the microdata file. If the probability that a specific record from the identification file corresponds to the same person as a specific record from the microdata set is high enough, then these two records are matched. This probability is the probability of re-identification per record, conditional on a particular disclosure scenario.

The results claimed by Paaß and Wauschkuhn are quite impressive. The number of correctly matched records was high, whereas the number of incorrectly matched ones was low. However, these results are based on a synthetic population. Starting with real data two data sets were constructed synthetically by adding noise to randomly chosen records of the real data set. The real data set consisted of both continuous data and categorical data. One of the two synthetically constructed data sets was the microdata file, the other one was the identification file. The kind of distribution of the added noise, with mean 0, is assumed to be known to the attacker. The variance of the noise, however, is assumed unknown to him. The attacker has to estimate this variance on the basis of the (assumed) knowledge of the statistical production process.

By applying their method Paaß and Wauschkuhn also showed that it is not necessary to make an accurate estimate of the unknown variance of the noise in order to obtain such good results. The results of Paaß and Wauschkuhn are based on *synthetically* generated data and this might be a weak point. Müller *et al.* (*cf.* [50]) applied the method recommended in [52] to *real* data. Their results are quite different from those of Paaß and Wauschkuhn. The percentage of correctly matched records was much lower in this study than in the study of Paaß and Wauschkuhn. Moreover, simple matching — *i.e.* a record is considered re-identified by an attacker if he succeeds in finding a value set, unique in the microdata file, which is *identical* to a value set, unique in the identification file — was not inferior to the method suggested by Paaß and Wauschkuhn. Apparently, the number of correctly matched records after application of the method by Paaß and Wauschkuhn was in disagreement with the probability of re-identification per record.

It seems that the method of Paaß and Wauschkuhn does not provide us with the desired probability of re-identification per record. However, the method does have its strong points, such as the idea of considering different scenarios, the assumption of measurements errors in the microdata file and the identification file, and the use of discriminant analysis.

In the context of masking procedures, *i.e.* procedures for microdata disclosure limitation by adding noise to the microdata, Fuller (*cf.* [37]) obtained an expression for the probability that a specific target is in the released data set. His approach has not been tested on real data. See [78] for comments on the approach by Fuller.

Paaß and Wauschkuhn, and Fuller are mainly interested in the effects of noise that has (unintentionally and intentionally, respectively) been added to the data. A weak point of their respective approaches is the, implicit, assumption that the key is a high-dimensional one. Assuming a high-dimensional key implies that (almost) everyone in the population is unique, so that the probability that a key value occurs more than once in the population is almost equal to zero. This makes the computation of the probability of re-identification per record considerably easier. On the other hand, in case of low-dimensional keys it is not unlikely that certain key values occur many times in the population. Therefore, deriving a probability of re-identification per record for low-dimensional keys is much harder than for high-dimensional keys.

Concluding the discussion above we can say that the idea to develop a working model yielding a re-identification risk per record appears to be overambitious at the moment. So, now we consider a less ambitious goal, namely a re-identification risk for an entire file. It is clear on the outset that such a model has rather limited use since it tends to accept records which are very easy to re-identify as soon as they appear in a large mass of other less "dangerous" records.

In a somewhat less ideal world a releaser of microdata would not be able to determine the risk of re-identification for each record. Instead he would be able to determine the risk that an unspecified record from the microdata set is re-identified. In this case, the statistical office should decide on the maximal risk it is willing to take when releasing a microdata set. If the actual risk is less than the maximal risk, then the microdata set can be released. If the actual risk is higher than the maximal risk, then the microdata set has to be modified. An example of such a model is the one proposed by Willenborg et al. (cf. [79]) and Mokken et al. (cf. [49]). This model takes three probabilities into account. The first probability, f, is equal to the sampling fraction. In other words, f is the probability that a randomly chosen person has been selected in the sample. The second probability, f_a, is the probability that a randomly chosen person is an acquaintance of a specific researcher who has access to the microdata. The third probability, f_u, is the probability that a randomly chosen person is unique in the population. The probability that a record from a microdata set is re-identified can be expressed in terms of these three probabilities, f, f_a and f_u (cf. Section 5.6).

Summarizing the main conclusions from the discussion so far: in practice none of the theoretical models seems to provide a suitable basis for SDC and we have to rely on heuristic arguments instead. In practice, rules for SDC of microdata could be based on testing whether scores on certain low-dimensional keys are "rare" or "common", i.e. occur frequently enough in the population. As SDC measures to be applied we consider global recoding and local suppression.

Applying this approach one faces a few problems that need to be solved: the determination of the keys that have to be examined, the way to estimate the number of persons in the population that score on a certain key, and to make operational the meaning of the phrase "frequently enough". In most cases the following rule is applied to overcome this last problem: a key value is considered safe for release if the frequency of this key value in the population is larger than a certain threshold value d_0. When applying this rule we are facing the problem that we do not know the number of times that a key value occurs in the population, if only sample information is available. Therefore, the population frequencies of key values should be estimated. Some estimation methods are examined in Section 5.2. In

Section 4.3.4 and Section 5.3 the special case of continuous variables is considered.

2.6   SDC for Tables

Between SDC for microdata and tables there are several similarities, such as the actions to produce safe data: global recoding and local suppression. In case of tables one refers to these actions as collapsing rows, columns *etc.*, table redesign (both for global recoding), and cell suppression (for local suppression). Also the general plan to apply table redesign and primary suppression to protect tables is similar to that of microdata: first identify the sensitive cells; if there are relatively many, then redesign the table, until only a few such cells are left; then suppress the cell values of any remaining sensitive cells. The final step, to be carried out if some sensitive cells have been suppressed and if marginal totals are presented as well, is to find some additional suppressed cells to protect the sensitive ones. This latter step is characteristic for tables although it can occur in microdata also [2]. The reason for this step is the presence of additional constraints as a result of the marginal tables. In fact, it is this secondary cell suppression problem that is, traditionally, at the heart of the SDC problem for tables. In fact, a more comprehensive setting of the core problem of SDC for tables is suggested by the analogy with SDC for microdata: it should include both table redesign and cell suppression.

The similarities between SDC for tables and that for microdata do not end here. The possibility of adding noise to the (continuous) values of sensitive variables is another similarity. We pointed at the difficulties one has to be prepared for when applying perturbation techniques to microdata, of which this one is an example (*cf.* Section 1.3.1). Such difficulties do not seem to exist when noise is added to values in tables. Usually *adding noise* to tables is done in a very orderly fashion: all values in the original table are rounded to one of the two nearest multiples of some well-chosen base value. This would be easy enough if no extra conditions were required. But, alas, the presence of marginal tables spoils the fun: usually one wants the rounded table to inherit the additivity property of the original table. That is, the rounded values in the interior of the table should add to the corresponding rounded marginal totals. This is called controlled rounding. With this requirement the rounding problem becomes non-trivial. It should be remarked that requiring the rounded table to be additive is not only a matter of cosmetics: there are examples showing that one can reconstruct

---

[2] for instance, when the profit, turn-over and the costs of an enterprise are given in a record: suppressing only one of these values is not sufficient, because it can be computed using the other two values.

the original table from the rounded one, if each cell value has been rounded independently of the other cell values. It can also be shown that there exist tables for which there is no rounded table satisfying all the conditions stated. More on rounding can be found in Sections 6.4.4 and 7.4.

Now we have discussed some similarities between SDC for tables and for microdata, the question is: What are the differences? First of all the "rare combination" criterion for microdata is different from any criterion usually applied to for tables. For tables with continuous data a sensitivity criterion such as a dominance rule is used (see below). Mechanically applying a dominance rule to a table with frequency count data would result in ·declaring the cells with values less than a certain threshold value as sensitive. This situation is very similar to the one in a microdata set when certain "rare" combinations occur in a microdata file that are considered unsafe.[3] Another difference with SDC for microdata is that standard SDC theory for tables assumes that the table contains information about the entire population and not about a sample. In fact, this issue seems to be almost totally neglected in the literature. It takes some extra work to modify the theory for tables with sample data. In the remainder of this section we will assume, however, that the tables are based on an observation of the entire population.

After these initial remarks on the similarities and dissimilarities between SDC for microdata and for tabular data we discuss some issues pertaining to tabular data in some more detail. The emphasis is on conceptual matters rather than on technicalities.

Prior to releasing a table one should check whether it contains any sensitive cells. This is usually done by means of a dominance rule. A dominance rule states that if the values of the data of a certain number of respondents, say 3, constitute more than a certain percentage, say 75%, of the total value of the cell, then this cell has to be suppressed. The main idea on which this approach is based is the following. If a cell is dominated by the value of one respondent, then his contribution can be estimated fairly accurately by the cell total. In particular, if there is only one respondent then his contribution can be disclosed exactly. If the value of a cell is dominated by the contributions of two respondents, then each of these respondents is able to estimate the value of the contribution of the other one accurately. In particular, if there are only two respondents then each respondent can disclose the contribution of the other one exactly. If there are m respondents then $m - 1$ of them, when pooling their information, can disclose information about the value of the data of the remaining respondent. For small m, say, 2, 3 and 4, this poses a problem. (The larger m, the less likely that a group of $m - 1$ of them will pool their information; cf. also Section 1.3.2.)

---

[3]This remark about frequency count tables is only meant to indicate a similarity; it is not intended to express that this should be the way to handle such tables when trying to make them safe for release.

Apart from dominance rules other rules for determining sensitive cells have been suggested in the literature. An example of such a rule is the prior-posterior rule (*cf.* [10]). This rule uses two parameters, p and q with $p < q$. It is assumed that every respondent can estimate the contribution of each other respondent to within q percent of its respective value. After a table has been published the information of the respondents changes and they may be able to make a better estimate of the contribution of another respondent. A cell is considered sensitive if it is possible to estimate the contribution of an individual respondent to that cell to within p percent of the original value. More on sensitivity measures can be found in Sections 6.2, 6.3 and 7.2.

Tables with sensitive cells have to be "treated" prior to their release. This treatment is aimed at removing the sensitive cells. In Section 1.3.2 two such techniques have been illustrated: tabular redesign and cell suppression. Redesigning the table, by combining suitably chosen rows or columns, *etc.*, is, however, not always a feasible technique, for instance if the tabular layout is fixed, as a consequence of previous publications of similar results.

The suppression of a cell because the contents of this cell is considered sensitive according to, for instance, a dominance rule is called *primary suppression*. Primary suppression alone is generally not sufficient to obtain a table which is safe for release. In a table the marginal totals are mostly given as well as the values of the cells. The value of a cell which has been suppressed can then often be re-computed by means of the marginal totals. Therefore, additional cells have to be suppressed in order to eliminate this possibility. This is called *secondary suppression*. Secondary suppression can be done in many different ways. Usually secondary suppression aims to optimize some target function, expressing "information loss". For instance, one could try to minimize the number of respondents whose data are suppressed in the table or one could try to minimize the total value of the data which are suppressed. Selecting the "best" target function is not straightforward and is generally based on subjective considerations. General guidelines as to make a motivated selection seem to be lacking.

Secondary suppression causes other problems as well. Although it might be impossible to re-compute the values of the suppressed cells in a table exactly after secondary suppression, it usually remains possible to compute the ranges in which the values of these cells lie, when it is known for instance that the values of the cells are all nonnegative. If these ranges of feasible values are small, then an attacker is able to obtain good estimates for the values in the suppressed cells. Therefore, secondary suppression must be done in such a way that the ranges in which the values of the suppressed cells lie are not too narrow. Some rather difficult methods based on mathematical programming have been proposed for secondary suppression (*cf.* [38] and [43]). The corresponding algorithms are time-consuming for relatively large tables. A simple heuristic, which we will call the hypercube method, has been proposed by Repsilber (*cf.* [60]). This method is less time-consuming and has the advantage that the algorithm is easy to implement. More on secondary suppression can be found in Sections 6.4.2, 6.4.3 and 7.3. Part of Section 7.3 is devoted to the hypercube method.

So far we have only considered single tables with their marginals. In the literature the dimension of these tables is generally two. Secondary cell suppression in higher-dimensional tables seems to be difficult. The setting in which to consider the SDC problem can, however, be generalized. This generalized setting is not only a natural one from a theoretical point of view, it also happens to occur in practice from time to time. The setting that is meant is that of *linked tables*, *i.e.* tables with common variables produced by different aggregations from the same microdata set. In fact the usual setting of a single table with marginals is a special case of a linked table setting. Consider for instance a 2-dimensional table with marginals. In this case we have three linked tables: the 2-dimensional "interior" table and two 1-dimensional tables. The situation for linked tables is, of course, more difficult than the usual setting of a single table with marginals. In practice the situation of linked tables is particularly relevant if the set of linked tables is fixed, but one has to decide on the exact design of each table and the cells to be suppressed. One should not rely on the theory of linked tables if it is not known which tables will be published from a given microdata set, a base file. In order to be sure that no unsafe tables will be released then, and avoiding that an extensive administration has to be kept about the tables released so far, it would be the most attractive strategy to make sure that the base file itself is safe (in the sense that any table produced from this base file would be automatically safe). The question, of course, is whether such a safe base file exists which contains still enough information to be of any interest. More on linked tables can be found in Section 7.5. But a lot more needs to be explored than that section brings to the fore.

We end this section with a warning, namely that one should, as a data disseminator, be careful about communicating the SDC measures one has used to protect a data set. In particular one should not say which parameter values have been used in these measures. To illustrate this in case of tabular data consider the following example. Suppose that the following dominance rule is used, and that this is publicly known: a cell is suppressed if at least 80% of the value is the combined result of the data of two companies. Suppose, furthermore, that the table contains information about all the companies in the population. Now, suppose that there are three companies contributing to a certain cell total and that the data of the largest company constitutes 50% of the value of this cell. If the cell is not suppressed, then this company can deduce that the data of the second largest company constitutes between 25% and 30% of the unpublished total value of the cell. On the other hand, if the cell value is suppressed then the largest company can deduce that the contribution of the second largest company constitutes between 30% and 50% of the total. In other words, the largest company can deduce that the contribution of the second largest company lies between 60% and 100% of its own value. Note that this result is obtained without using any extra information in the table! Absolute secrecy about the parameter values of the dominance rule applied would have prevented these deductions.

3
Policies and Case Studies

The policy of adapting one's self to the circumstances
makes all ways smooth.
—Lavater

3.1 Introduction

In this chapter an overview of the SDC policies for microdata of various national statistical offices is given. Moreover, a few case studies are presented. The focus is on microdata rather than on tabular data, because SDC policy with respect to microdata is more controversial than with respect to tabular data.

Before we begin by discussing the policies and case studies in detail we first describe the various options for a statistical office to release its data in Section 3.2. In Section 3.3 we describe the policies of national statistical offices with respect to SDC for microdata. An institution, the WSA, has been founded quite recently in the Netherlands to facilitate the dissemination of microdata sets. This institution serves as an intermediary between Statistics Netherlands and the users of microdata sets issued under license. The role of the WSA in the distribution of microdata sets issued under license is examined in Section 3.4. The second case study comes from Great Britain and concerns the release of Samples of Anonymized Records (SARs), which are produced from GB Census data. Special attention is given to the protection of confidentiality in the SARs. The SARs are examined in Section 3.5. The policy on SDC of Eurostat, the Statistical Office of the European Communities, deserves special attention. This statistical office does not carry out surveys itself. For its data Eurostat depends on its member states. These member states frequently release their data to Eurostat only after these data have been protected against disclosure. The

rules and techniques by which these data are protected depend on the particular member state. It is clear that this implies that data collection may
present a lot of problem for Eurostat. Also when Eurostat wants to publish
data problems may arise, because some member states may consider their
contribution to be confidential. More details on Eurostat can be found in
Section 3.6. The final case study concerns the Luxemburg Income Study
(LIS), which is interesting because it is an example of on-line access to
data, *i.e.* remote access via modems and telephone lines.

3.2   Options for Data Dissemination

Statistical offices can release their data in several ways, ranging from small
tables containing little information to large microdata sets containing much
information. Traditionally, releasing data in the form of relatively small tables has been the favored way. Recently, mainly due to the increasing power
of PC's, (very) detailed tables and microdata sets have become important
products of statistical offices. A drawback of releasing very detailed information is, however, that the privacy of the respondents may be greatly
endangered. On the other hand, offering full protection against disclosure
of information of individual respondents will result in much less detailed
information. As a consequence, future users of the data will be curtailed in
performing their plans of research with the data. For some future users of
the data, especially researchers, this would present a serious problem.

In order to overcome these opposing demands one might try to find a legal
setting to solve the problem. Under the appropriate legal arrangements it
would not be necessary to reduce the disclosure risk to zero. A certain
disclosure risk in the data is then acceptable, provided that the user has
agreed to a particular code of conduct, stipulated in a contract between the
statistical office and the user. The legal setting is supposed to reduce the
possible misuse of the data, as a result of accidental re-identifications and
disclosures. Of course, this alternative automatically restricts the group of
potential users to those who are considered respectable researchers, deemed
trustworthy for sharing information that is not entirely free of disclosure
risks.

There are several degrees in the extent to which the requirement of safeguarding confidentiality might be loosened. In its extreme form it allows
access to the microdata in their original version. Evidently, this opportunity is open for a very restricted group of users. Moreover, these users
usually have to perform their research on-site, *i.e.* on the premises of the
statistical office, under special conditions (*e.g.* working in a special room
that is supervised, working on a stand-alone computer or a closed network
with locked floppy drives). Such users are treated in fact as if they were
staff members of the statistical office. That is, they have to comply with
the same codes of conduct as any staff member of the office; in particular

they are sworn to secrecy.

On the other end of the spectrum there is the possibility of releasing data without any obligation whatsoever from the part of the user with regard to proper use and confidentiality. This can be justified only if the disclosure risk for these data is virtually nil. In other words: SDC should be perfect in a technical sense. A consequence of this is that the usefulness of the data for many scientific purposes is also minimal. However, such publicly available tables and microdata sets may still be very useful for non-scientific use. For example, public use microdata files can be used for educational purposes, *e.g.* to train students in the application of multivariate statistical techniques on "real" data.

Tables are nowadays released in various ways: on paper, on floppy disks, on CD-Roms, on tapes, *et cetera*. Also releasing them via on-line access or via the Internet is possible. When tables are released via in this way, then users have to connect their computer to the computer of the statistical office by telephone or via an Internet connection. Using this connection they can download tables from the office's computer to their own computer. There are basically two possibilities for downloading of tables. Firstly, everybody has access to the tabular data. In that case only standard tables with a relatively low level of information content can be offered, in order to restrict the disclosure risk. Secondly, only a select group of persons, namely researchers that are deemed trustworthy, have access to the data. In that case, one may allow these users to specify their own tables (under certain restrictions). The information content of the tables can therefore be relatively high. In this second case the users first should sign a contract. For a practical example of on-line data access see Section 3.7, which contains information on the Luxemburg Income Study. In Table 3.1 the options for data release are summarized.

As can be seen from Table 3.1 statistical offices can disseminate microdata in basically three ways: with low information content and without a license, with high information content and with a license and with very high information content, in which case it is usually required for the researchers to work on-site.

3.3   SDC Policy for Microdata of Various Statistical Offices

In order to obtain a more detailed picture of the policy of the various national statistical offices with respect to SDC for microdata Citteur and Willenborg carried out a survey among several of them in 1991. Their findings are summarized in [8]. As was to be expected, there was a considerable variation in SDC policy among the various statistical offices. The main findings of this SDC policy survey are presented below, country-wise

TABLE 3.1. Options for data dissemination.

Dissemination options	Information Content	License	On-site
Tables for public use	low	no	no
Tables on request (ordinary)	low	no	no
Tables on request (special)	high	yes	sometimes
Tables on-line (ordinary)	low	no	no
Tables on-line (special)	moderately	yes	no
Microdata for widespread use (public use files)	low	no	no
Microdata for research (under license)	high	yes	no
Lightly protected microdata	very high	yes	usually

in alphabetical order. It should be stressed that the situation described at a particular statistical office may have changed in the mean time.

Australia — Australian Bureau of Statistics

For a number of surveys public use microdata are available, provided that the Australian Statistician, who is advised by the Microdata Review Panel, approves such release. Well-known techniques such as deletion of variables and collapsing of categories, including the formation of geographic units containing 200,000 inhabitants at least, are applied. Where this facility is inadequate to meet user demands the Australian Bureau of Statistics may consider running the user's computer program on the data and releasing the results instead of the individual data themselves.

Canada — Statistics Canada

SDC policy resembles that of Australia. The screening of microdata prior to their release is the responsibility of the Microdata Release Committee.

Denmark — Danmarks Statistik

There are no public use microdata at all. In order to compensate for this there is the possibility to apply for microdata on the basis of a license, or to have access to them on-site.

France — Institut National de la Statistique et des Études Économiques

Two forms of data access exist: public and on the basis of a license. In the latter case, also data on enterprises are accessible, provided that the Conseil National de l'Information Statistique gives its consent.

German Federal Republic — Statistisches Bundesamt

No public use data are available at the micro-level. Release on the basis of a license is possible. As a special instance we mention the facility for statistical services of municipalities, which are allowed access to the nearly unmodified Census-data, as far as their own population is concerned. Otherwise, access to survey-data on the basis of a license is the exception rather than the rule. Moreover, several protective measures accompany such a release. The strongest of them are out-dating (the data should have been collected sufficiently long ago), subsampling and collapsing of regions until these contain 500,000 inhabitants at least. Due to the demands by the scientific community for a less restrictive policy extensive research has been conducted (cf. [5] and [53]) on the risks of disclosure from microdata. Both synthetic and existing files were used in these investigations. Various scenarios about the kind of user of the data were assumed, such as that of a journalist or a detective. In this context the profits, arising from a potential disclosure, were contrasted with the costs — often thought to be substantial — in accomplishing a disclosure, as well as the "quality" of the disclosed information, in terms of accuracy and validity.

Great Britain — Office of Population Censuses and Surveys

Public use data are available for various surveys. The most important technique for protecting confidentiality is collapsing geographical codes. After extensive discussions and preparations two standard files from the 1991 Census, the so-called Samples of Anonymized Records (SARs), have been formed and made available to academic and non-academic users alike. Both types of users have to sign legally binding agreements. As in the Netherlands and Norway, there is a central institution which issues the SARs. More about the SARs can be found in Section 3.5.

Italy — Istituto Nazionale di Statistica

As in Germany access to microdata is rather restricted. Local authorities are entitled to use Census data in their original form, as far as their own population is concerned. In order to create more facilities research was carried out on existing microdata from the Census and the Causes of Death survey. Superpopulation models are used in evaluating estimates of the number of population uniques with respect to various combinations of identifying variables.

The Netherlands — Statistics Netherlands

An explicit distinction is made between public use files and microdata issued under license. For both types of files general SDC rules have been formulated. The policy is to encourage the use of standard microdata files (of either type), by putting price tags on non-standard products.

Public use files are protected purely by adapting the original data file (through global recoding and local suppression). A public use file should virtually be free of any disclosure risk. Thus it is not allowed to specify in a public use file a region by direct indication. Only regional characteristics — such as by degree of urbanization — are allowed, provided they define only sufficiently diffuse areas. The number of identifying variables should be limited. Each category of an identifying variable should occur "frequently enough" in the population. Likewise, bivariate combinations should occur "frequently enough" in the population. Moreover, really sensitive information on e.g. sexual behavior, having a criminal record, ethnicity, etc. must be absent from these files. The data should have been collected at least one year prior to their release.

In contrast, the regime for microdata issued under license is far less severe. A region may be mentioned explicitly, provided it is not too small (in terms of the number of inhabitants). There is a trade-off between the specification of regional detail and the degree of detail by which the type of employer, the occupation and the education of a respondent is specified. There is no restriction on the number of identifying variables. Scores on certain combinations of identifying variables should occur "frequently enough" in the population.

A new facility, the so-called WSA, was created in 1994 in order to enhance the release of microdata issued under license. More about it can be found in Section 3.4.

New Zealand — Department of Statistics

Files with individual data are not available on a large scale. Two instances can be mentioned: a data file which is derived from a survey on income and expenditures by households and data from the Census. The former file is released to government agencies only and under strict conditions of use. The Census data are released in the form of aggregates, for which one can choose between various alternatives with respect to information about regions, ranging from broad to fine (100 persons in the population). As one is provided with more regional detail, other information is limited, e.g. by collapsing categories, deletion of variables and rounding. Data-release should be approved by the Government Statistician.

Norway — Statistisk Sentralbyrå

No public use data are available. For Academia and Government however there is the possibility to obtain access to individual data from several surveys on the basis of a license. Like in the Netherlands the transfer of such information takes place via a central organization, the Norwegian Social Science Data Service. It is to be noted that some identifying variables are given in considerable detail: age in years, occupation in a four-digit code and education in a five-digit code. In contrast, code of region is collapsed from four digits to one only. In addition, outdating is considered as a useful means in SDC. As in the case of Australia, computations can be carried out on the original data, at the request of researchers.

Sweden — Statistics Sweden

No public use data are available. Access on the basis of a license and also on-site is possible. As a special feature the on-line facility should be mentioned, which, however, is restricted only to the production of tables. Like in Norway, "old" data are released more quickly than recent ones.

United States — U.S. Bureau of the Census

Public use data is the only way to release microdata to outside users is the general philosophy. So there are no microdata for special categories of users, as e.g. in the Netherlands. However, permission might be granted to researchers to work on-site, on the original data. Mostly, this facility is open to persons who are cooperating with the Bureau of the Census in a given project, are performing a project, related to the data, for the Bureau or are entitled by law to investigate the activities of the Bureau. In the case of the public use data any region that can be identified should have at least 100,000 inhabitants. But if deemed necessary, for instance in case of longitudinal data, a higher limit can be set. Special attention is paid to prevent linking Census Bureau files to other, outside files. Aggregate data may be formed on request, including variance-covariance matrices. A Microdata Review Panel screens the data on disclosure risks before release. Like in other countries for a given survey only one (standard) file is available.

With this description of the policy of the Bureau of the Census with respect to SDC we have come at the end of this section. We can conclude that the policies with respect to microdata can be quite different: some offices release no microdata sets at all, others release only microdata with a license, still others release only public use data and the remaining ones release both microdata under license as public use microdata. In the next sections we will examine some special cases, such as, for instance, the WSA in the Netherlands and the SARs in Great Britain.

3.4   The Netherlands: WSA

Statistics Netherlands releases data in several ways. Tables are published on paper and on floppies. It is also possible to have on-line access to certain (standard) tables. Microdata sets are released in three ways: as public use files (without a license), as microdata sets for research (with a license) and via on-site access. In order to facilitate the release of microdata for researchers a special agency was established (in the spring of 1994): the *Wetenschappelijk Statistisch Agentschap* (WSA), *i.e.* as the *Scientific Statistical Agency*. The WSA is part of the Netherlands Scientific Research Organization.

The motivation for founding the WSA came from the fact that negotiations between Statistics Netherlands and the scientific community on the contents (the variables, their level of detail, *etc.*) of microdata sets quite often were troublesome and time- (and hence) money-consuming. This was mainly caused by the tailor-made character of each release, whereby each time again the entire data protection "ritual" had to be performed. Moreover, in many cases, despite considerable efforts, the resulting files did not really satisfy the user. Yielding to the demands of the user, however, would often lead to files with unacceptably high disclosure risks. It was clear that this controversy would never stop, unless a more radical step in the relation between Statistics Netherlands and the scientific user community would be taken. At the initiative of the Minister of Education and Science, Statistics Netherlands and the scientific users community started talks about a new organizational structure for dissemination of statistical data that would be acceptable for both parties. The aim was, as a matter of fact, not only to agree upon a better and more efficient dissemination procedure. Another aim was to find a way to reduce the price the users have to pay for the data considerably, without, however, the implication of any negative financial consequences for Statistics Netherlands. First a pilot project was started in which external users, in close cooperation with Statistics Netherlands, constructed a safe (according to the rules of Statistics Netherlands) microdata file for the Dutch Continuous Labor Force Survey. For the users involved this was an instructive project: they could experience themselves that producing a safe data set (according to the rules of Statistics Netherlands) is a rather time-consuming, error-prone and tedious craft. This also created an opportunity for the users to appreciate the arguments of Statistics Netherlands for its restrictive SDC policy.

A consequence of this pilot project was that an institution should be set up in which both parties should participate in reaching compromises for particular data releases. This idea resulted in establishing the WSA. A main feature of this agency is its responsibility to channel the wishes and demands of the user community, as well as of Statistics Netherlands with respect to the formation of standard files. The operating constraint is that there should be at most one standard research file for a survey conducted. Each standard file to be issued should be deliverable "off the shelf", complete with a code book and other relevant meta-information. The Netherlands Scientific Research Organization, WSA's host, and Statistics Netherlands agreed that Statistics Netherlands should transfer files of at least eight surveys every year. As an annual reimbursement Statistics Netherlands receives a certain amount of money from WSA, equaling the amount that Statistics Netherlands is annually supposed to generate from selling its microdata. This agreement has a term which ends after four years in 1998; an evaluation will then take place.

Scientists who want to be eligible for access to this facility should be affiliated with a university or with an institution related directly to a university, or with another institution performing pure or policy oriented research. In the code of conduct to which a user should adhere (stipulated in a license), important conditions are, amongst others:

- Data should be used for statistical analysis only, administrative applications directed at the individual are excluded.

- No information assignable to a well-specified unit should be transparent from the results to be published.

- Before the results are published they should be submitted to Statistics Netherlands in order that it can check if confidentiality is safeguarded.

- Matching to other files is prohibited with the sole exception in case the Director-General of Statistics Netherlands has given his permission.

- Transferring data to other users is prohibited.

Further conditions are concerned with aspects of:

- *Property*: ownership of the data remains with Statistics Netherlands.

- *Term of the contract*: the data should be destroyed after three years at most; this destruction should be affirmed in writing to Statistics Netherlands.

- *Storage*: data should be stored on a safe computer system, thereby excluding open networks to which persons other than those who are entitled and mentioned in the contract have access. Making sure that only those who are mentioned in the contract have access to the data belongs to the responsibility not only of the user, but also of the institution to which the user is affiliated.

The last of these conditions illustrates that responsibility extends to the institution of the user. This aspect is apparent also from the fact that it is the institution which has to sign a contract with Statistics Netherlands in which these conditions are included. In a separate contract the user himself pledges to observe the obligations of the former one. In addition he has to certify that all information about individual units that might come to his knowledge through working with the data will be kept secret by him and, again, that he will not publish any result from his analysis prior to written permission by Statistics Netherlands.

In its first year of existence files of 15 surveys have been transferred to WSA. They cover a broad range of subjects of social research; even one survey on an economic field is included. It should be noted that the composition of the bundle of files available should be decided upon every year. Also the choice and detail of the variables to be included in the files pertaining to a particular survey might vary from year to year. Table 3.2 on page 39 lists the microdata sets currently available through WSA.

3.5 Great Britain: Samples of Anonymized Records

3.5.1 Samples of Anonymized Records

For the first time in a British census, the 1991 statistical output included Samples of Anonymized Records (SARs). Known as Census Microdata or Public Use Sample Tapes in other countries, SARs differ from traditional census output of tables of aggregated information in that abstracts of individual records are released. The released records do not conflict with the confidentiality assurances given when collecting census information, since they contain neither names or addresses nor any other direct information which would lead to the re-identification of an individual or household. Essentially three per cent of the records have been released in two samples. The SARs offer users the freedom to import individual-level census records into their own computing environment and the ability to produce their own tables or run analyses which are not possible using aggregated statistics.

Details of the SARs

Two SARs have been extracted from the GB censuses:

1. A two per cent sample of individuals in households and communal establishments.

2. A one per cent hierarchical sample of households and individuals in those households.

TABLE 3.2. Microdata files of Statistics Netherlands currently available through WSA

File	Period [a]	Vars	Size # Units	Price [b]
Career				
Labor Force Survey	1	60	84,000	5,000
Socio-Economic Panel Survey	1	500	14,000	5,000
Educational Cohort Study	5	120	20,000	3,000
Family Expenditure Survey	1	900	2,800	3,000
Family Planning Survey	5	180	6,000	2,000
Consumer Confidence Survey	1	30	12,000	1,000
Living condition				
Housing Needs Survey	4	570	54,000	5,000
Health Survey		630	9,000	5,000
Day Recreation Survey [c]	1			4,000
data on persons		20	25,000	
data on day trips		50	59,000	
Life Situation Survey	1	520	6,000	4,000
Victimization Survey [d]	1			2,000
data on persons		150	4,400	
data on victimizations		190	2,200	
National Election Survey	4	400	1,800	1,000
Mobility				
Mobility Behavior Survey	1	70	21,000	2,000
Car Panel Survey	1	50	13,000	1,000
Building				
Building Permits Issued	1	20	45,000	1,000

[a]periodicity: 1 = yearly or continuously, 4 = once every four years, 5 = once every five years.

[b]price of recent edition in Dutch guilders (Dfl.) for academic users. Historic files are less expensive, non-academic users are charged higher prices 1 Dfl. ≈ \$ 0.60, 1 Dfl. ≈ 0.45 ECU. Subscription to the complete bundle is possible at a price of Dfl. 25,000.

[c]the Day Recreation Survey consists of two separate files.

[d]the Victimization Survey consists of two separate files.

The two per cent SAR has finer geographical detail and the one per cent SAR has finer detail on other variables. The two per cent individual SAR contains some 1.12 million individual records (1 in 50 sample of the whole population enumerated in the census). It was selected from the base which lists persons at their place of enumeration. Details are given as to whether or not the person was a usual resident of that household, and if so (and enumerated in a household) whether they were present or absent on census night. The following other information is given for each sampled individual:

- Details about the individual ranging from their age and sex to their employment status, occupation and social class.

- Details about the accommodation in which the person is enumerated (such as the availability of a bath/shower and the tenure of the accommodation) or, if they were in a communal establishment, the establishment type (hotel, hospital, *etc.*).

- Information about the sex, economic position (in employment, unemployed, *etc.*), and social class of the individual's family head.

- Limited information about other members of the individual's household (such as the number of persons with long-term illness and numbers of pensioners).

In effect, all the census topic variables listed are on the file; the only exceptions are variables either suppressed or recoded to maintain the confidentiality of the data; these are listed below. In all, there are about forty pieces of information about each individual. The one per cent household SAR contains some 240,000 household records together with sub-records, one for each person in the selected household. Information is available about the household's accommodation together with information (similar to the two per cent sample) about each individual in the household and how they are related to the head of the household.

3.5.2 Confidentiality Protection in the SARs

The census offices in some European countries have refused to release microdata because they believe, on the basis of research such as that conducted by Paaß(*cf.* [53]) and Bethlehem *et al.* (*cf.* [3]), that the risks of disclosing information about respondents' identities are too high. Much of this work is concerned with how many people have unique combinations of census characteristics which would make them open to identification. The Economic and Social Research Council Working Party which negotiated the release of the SARs took the view that uniqueness was only one part of a four-stage process of disclosure: data in the microdata file would have to be recorded in a way compatible to that in an outside file, the individual in an outside file would have to turn up in a SAR, the individual would have to have unique values on a set of key census variables and it would be necessary to verify that this individual was also unique in the population. Rough estimates of the size of risk at each stage were made; when cumulated, the risks of disclosure appeared very low; multiplying the various probabilities together, the working party concluded that the risk of anyone in the population being identifiable from their SAR record were extremely remote; their best estimate was something of the order of 1 in 4 million. (For more details of such calculations, consult [47], [48], [65] and [68]). The arguments put forward were important in persuading the census offices to release the SARs suitably modified to protect anonymity where this was felt at risk. In this section the various disclosure protection measures taken are described.

Sampling as Protection

The low sampling fractions of the SARs offer a strong source of disclosure protection for sensitive data. It not only reduces the actual risk that a particular individual can be found in the census output, but it probably has its greatest effect by reducing the chances that anyone would make the attempt at identification by this means. The two SARs (a one per cent sample of households and a two per cent sample of individuals) are sufficiently small to offer a great deal of protection; the samples do not overlap, so that the detailed household or occupational information available on the household file cannot be matched with the detailed geographical information available on the individual file.

Restricting Geographical Information

One of the key considerations which may affect the possibility of disclosure of information about an identifiable individual or household is the geographical level to be released (*i.e.* how much detail is given about where the person was enumerated). The full census database holds information at enumeration district (ED) level (about 200 households or 500 persons in each ED) and even at unit post-code level (about 15 households). If released, such detailed geography would obviously pose a confidentiality risk. Empirical work and comparisons with SARs released in other countries showed that a sensible level for the individual (2%) SAR to be released would be areas equivalent to large local authority districts.

To be separately identifiable, the decision was taken that an area had to have a population size of at least 120,000 in the mid-1989 estimates. The primary units used were local districts; only one geographical scheme was permitted, or smaller areas could be identified in the overlap, say between a local district and a health district. A population size of 120,000 is slightly higher than the lowest level of geography permitted in the US SARs (100,000), but it still has the advantage of allowing all non-metropolitan counties in England and Wales, most Scottish regions, all London boroughs (except the City of London), and all metropolitan districts to be separately identified. Smaller local authority districts (under 120,000 population) were recoded to form areas over 120,000, with contiguity and similarity of socio-economic characteristics used as criteria for making decisions over global recoding.

The one per cent household SAR, because of its hierarchical nature (*i.e.* statistics about the household and all its members), is more of a disclosure risk. For this reason it was decided that, for this SAR, the lowest geographical detail revealed would be the Registrar General's Standard Regions, plus Wales and Scotland. The only exception is that the South East is split into Inner London, Outer London, and the Rest of the South East Region.

It should be noted that the order of records in both SARs has been re-arranged before the Census Offices released them. This is to prevent any possible tracing of individuals or households back through a region or district.

Suppression of Data and Recoding of Categories

Some alterations have been made to the data to reduce the number of rare and possibly unique cases. Information which is unique in itself, such as names and addresses, has been omitted altogether; (technically these variables have not been suppressed since they are never put on the computer). Precise day and month of birth have been suppressed.

The Thresholding Rule

The degree of detail permitted on other variables was the subject of a thresholding rule which ensured that the expected value of any category at the lowest level of geography on any file was at least 1. The threshold, when operationalized, dictated that a category must have 25,000 cases in it in the GB file before it could be released on the individual SAR, or 2,700 cases before it could be released on the household SAR.

With some other variables, the smaller categories have been globally re-coded, either across the entire range of the variable or only at the extremes, *i.e.* by a process known as top-coding. The rule used to decide the level of detail to be released was to globally recode categories to a sufficient degree, so that, on average, the expected sample count would be at least one for each category of each piece of information for the lowest geographical area permitted on each SAR.

Some justification for restricting attention to the distribution of the univariate categories of each variable in turn was given by Marsh *et al.* (*cf.* [48]). They demonstrated that the risk of an individual having a unique combination of values of a set of variables could be predicted with a high degree of certainty simply from knowledge of their membership of rare categories of each variable taken singly. The precise cut-off at an expected value of 1 in the file was set at a value sufficiently high to give a reasonable protection of anonymity. The rule was applied to each census variable. Expected counts were obtained by using 1981 Census frequency counts (supplemented by more recent surveys, for example the Labor Force Survey) at the national level for the whole population. To obtain expected counts, the count of 1 per category per SAR area was grossed up to the national level:

$$C = 1/f \times (N/S),$$

where

C = expected count at the national level.

f = sampling fraction (1/50 for individual SAR and 1/100 for household SAR).

N = national population (56 million).

S = smallest geographical area population (120,000 for individual SAR and 2.1 million (East Anglia) for household SAR.

Thus 25,000 and 2,700 were the two thresholds used for the individual and household SARs respectively. In theory, a small amount of random noise could have been added to certain variables in a manner analogous to the procedure adopted for the small area statistics. A technique similar to this has been used in the 1990 US Census for example: geography has been subject to a degree of perturbation by switching a small number of similar households between nearby areas (*cf.* [51]). However, the natural levels of noise in the data, combined with the analytical difficulties of minimizing bias to both measures of location and spread by such techniques in a multipurpose file led to perturbation not being implemented in any form for the SARs.

Global Recoding

When expected frequency counts fell below the threshold, categories were globally recoded. With some variables, recoding was only required at one end of the distribution: thus rooms were top-coded above 14 and the number of persons in the household was top-coded above 12. Two variables were both recoded and top-coded; with age, 91 and 92 were recoded, 93 and 94 were recoded and 95 and over was top-coded; with hours of work, 71–80 hours per week has been recoded and the rest top-coded above 81.

When variables were not measured on a numeric scale, judgments had to be made about which categories to be put together. Classifications for census data are often hierarchical. For example, for the Standard Occupational Classification there are 371 unit groups, 77 minor groups, 22 sub-major groups, and 9 major groups. In cases such as these, small categories could be amalgamated to the next level in the hierarchy. In other cases, detailed advice was sought from subject experts about how the groups should be formed. In the case of three variables in the two per cent individual SAR, it was deemed necessary to further recode categories, even though they contained numbers which fell above the threshold: occupation, industry, and subject of qualification. As a result of advice received from the Technical Assessor, occupation was reduced from the 220 categories proposed (out of a possible 371) to 73; similarly industry was cut from a possible 334 to 60 and subject of educational qualification from a possible 108 to 35. Almost full occupational detail remains on the one per cent household SAR, however.

There were other factors which determined the detail to be released:

- The categories of occupations and industries in the public eye were recoded further than mathematically necessary to guard against disclosure; *e.g.* actors/actresses and professional sportsmen/women.

- Large households were seen as a disclosure risk in the household sample. Applying the frequency rule to size of household, a large household in the 1981 Census was estimated to be one of 12 persons or more. Consequently, only housing information is given for households containing 12 or more persons. No information about the individuals in the household is given.

- Geographical information for such items as workplace and migration (address one year before census) has been heavily recoded. This is because of the high likelihood of uniqueness of such information when used in conjunction with area of residence.

3.5.3   Dissemination of SARs

The licensing and distribution of the SARs is the responsibility of Manchester University who have a contract with ESRC. The SARs may be used for both academic and non-academic purposes. All Higher Education Institutions (HEI) are required to sign an End User License Agreement which makes the HEI responsible for those members of their institution who are using the data. Users within each institution must be either members of staff or students and must sign a further individual registration form which contains a binding undertaking to respect the confidentiality of the data. Specifically, users have to guarantee not to use the SARs to attempt to obtain or derive information about an identified individual or household, nor to claim to have obtained such information. Furthermore, they have to undertake not to pass on copies of the raw data to unregistered users, and the Census Microdata Unit has the responsibility of auditing their use of the data. They must sign a statement that they understand that the consequences of any breach of the regulations on the part of any user in a specific institution can lead to the withdrawal of all copies of the data from that institution. Non-academic organizations sign a similar End User License Agreement and undertake not to allow the data to be used other than by their employees. Note the similarity of this End User License Agreement and the contracts for microdata sets that are used in The Netherlands (*cf.* Sections 3.3 and 3.4).

The data are free for the purposes of academic research; to get the data free the researcher must be doing the research in an institution qualified to receive an ESRC award, and the research must be funded either by the Universities Funding Council or one of the Research Councils. When the data is used either by those outside the academic sector or by researchers in universities for sponsored research, a charge is made for the data. In order to encourage a high volume of usage of a product whose advantages may not yet be well appreciated in Britain, these charges are being kept extremely low; an entire national SAR can be bought for 1,000 GBP + VAT, and subsets of a county or local district for 500 GBP. SARs are available on a similar basis for Northern Ireland, but, for non-academic users the cost is less.

3.6 Eurostat

The position of Eurostat differs from that of the national statistical institutes (NSI's) in that it does not perform surveys itself; it is dependent on the member states for the availability of microdata. In part these member states are under an obligation to provide their microdata to Eurostat. Eurostat compiles the information obtained into statistics on a European-wide level, often directed at a comparison between countries. Notwithstanding the legal obligation of providing their microdata, NSI's at some occasions had a certain reluctance in transferring their data to Eurostat, except after severe modifications of the data to reduce disclosure risk. In fact Eurostat was viewed by many NSI's as an external user. As a consequence, Eurostat felt itself unable to perform its duty.

A *Council Regulation on the transmission of data subject to statistical confidentiality to the Statistical Office of the European Communities* (June 11, 1990) is intended to remove any obstacles impeding the transfer of data files from European NSI's to Eurostat. Ample consideration is given to the "infrastructure" by means of which the confidentiality of the data has to be safeguarded. This also includes the transport from the NSI to Eurostat Special security areas and the creation of special computing facilities Eurostat within which all operations on the transmitted material should take place. Encryption techniques are used at various occasions. Correspondingly, access to these areas, and hence to the data, is allowed to specifically dedicated persons only. On the one hand the number of these individuals should be kept as small as possible. On the other hand, the "Four-eyes principle" is in force, by which the agreement of at least two persons is required before access is granted. All activities with respect to the data are audited. As an example we mention the automatic logging of all critical operations on the protected machines. The results, e.g. in the form of tables containing aggregates and frequency distributions, which are compiled from the microdata should be cleared by the member states involved, prior to actual publication. The various rules for maintaining confidentiality in statistical publications, which may differ substantially between the various states, have been taken into account already. Eurostat explicitly accepts the responsibility of adhering to these rules. As a result of the policy sketched, and the conditions created, NSI's are supposed to have no argument any longer for not withholding any data Eurostat is entitled to.

Eurostat faces some special problems as far as SDC is concerned. For instance, suppose that Eurostat wants to publish a table containing data from its member states, including marginal, in particular European, totals. Suppose furthermore that for reasons of confidentiality a member state does not want its contribution to this table to be published. (It should be noted that an NSI has a right to do so.) So, the contribution of this member state is suppressed. However, due to the presence of marginal totals in the table, this contribution can be re-computed. So, in order to preserve confidentiality there are several possibilities: the contribution of (at least) one other member state to the table has to be suppressed, the table should be redesigned such that the resulting table does not have any sensitive cells, or the marginal totals should not be published. The latter option is not very attractive to Eurostat, because a European statistical office should at least publish European totals. The first option leads to a rather difficult practical problem: at least one NSI should be found that is willing to suppress its contribution to the table, not for its own sake but out of solidarity with a "fellow" NSI. This implies that this contribution should not be derivable from any publication of the solidary NSI's themselves. So, suppressing the contribution of a member state to the table may have a considerable influence on its publications. Most member states will therefore be rather reluctant in cooperating in this secondary suppression process. The second option, i.e. that of redesigning the table, may seem the only feasible one in practice, although it may render tables with insufficient detail.

3.7   Luxemburg Income Study

A special instance of creating facilities for scientific research in cases where for reasons of privacy access to microdata with sufficient detail could not be afforded, is the Luxemburg Income Study (LIS). LIS is a databank with socio-economic microdata from over 20 countries and over 45 surveys. It is managed, as its name indicates, in Luxemburg. Many European countries contribute to LIS, and in addition Canada, the USA and Australia are participants. The main feature of LIS is that it performs the analysis according to the set-up that the external investigator communicates on-line with LIS. In other words, these users have no direct access to the microdata themselves. This drawback for the user is compensated by the fact that the plans envisaged by him are executed on the unmodified microdata. The output from this operation is returned to the user after a number of checks on the safety of these results has been made. In particular, no data on individual households should be recognizable from the output. Furthermore, it is obviously not allowed that the (SSPS) set-up submitted by the user contains commands such as *list cases, write cases* and the like, which would disclose individual records. A user has to sign a *LIS Pledge of Confidentiality*; a main part of it is the promise of using information from LIS for scientific purposes only. The communication with LIS, from all over the world, takes place via electronic mail.

LIS dates back to 1983. Publications resulting from it are numerous; up to 1994 they include more than 100 LIS working papers. In addition, workshops are organized to acquaint future users with LIS and to make students familiar with performing socio-economic research. LIS-facilities are open to members of a number of scientific and similar non-commercial institutions. LIS meets a real demand, as is witnessed by the intensity by which new jobs are submitted: approximately 100 a day. This large instream precludes that the jobs are stored for a long time; a storage time of one month is standard. When the jobs are removed, they are once more inspected on the basis of a sample in order to inspect at least the kind of information requested by the users. More information on LIS can be found in [46].

4
Microdata

Individuals, not stations, ornament society.
—Gladstone

4.1 Introduction

In this chapter we examine various aspects of SDC for microdata sets. These aspects could play a role in case a data disseminator would like to develop a systematic data dissemination policy, including types of data to be disseminated and general SDC rules defining what conditions a safe type of data would have to satisfy. The aim is not so much to advocate particular SDC rules, complete with choices for parameter values, but rather to point at various aspects that have to be taken into account. Therefore there is first a discussion of relevant aspects, followed by two concrete examples of SDC rules to give the discussion a focus. The rules shown here are derived from those that Statistics Netherlands currently uses, be it that the choices of the parameter values may be different. Not only the example SDC rules are strongly inspired by the current practice at Statistics Netherlands; the same holds true for the entire chapter. A description of the contents of the remainder of the present chapter follows.

In Section 4.2 several ingredients of SDC rules are examined. First of all there is a discussion of which variables are considered to be identifying. Especially, those variables of which (some of) the values are very visible (*i.e.* pertain to a characteristic that has this quality) and that can be used to locate the corresponding person have to be considered identifying. Examples of identifying variables are *Residence, Sex, Occupation* and *Education*. In Section 4.2.1 the classification of variables is discussed. SDC rules should describe which combinations of values of identifying variables need to be examined. The frequency of such a combination should be higher than a certain threshold value, in which case it is called *common*, and may be published. On the other hand if the frequency is below such a threshold value, in which case it is called *rare*, the combination is considered unsafe and SDC measures should be taken to eliminate it. The threshold values are parameter values to be specified by SDC rules. In Section 4.2.2 rareness criteria are examined. Some kinds of variables should be given special consideration by the SDC rules. Examples of such kinds of variables are sensitive variables, household variables, regional variables and sampling weights. Some sensitive variables, such as variables on sexual or criminal behavior, should be treated with more care than others. The SDC rules should describe the additional steps that should be taken to protect these variables sufficiently. The protection of sensitive variables is discussed in Section 4.2.3. Another special kind of variables are the so-called household variables. A household variable is a variable for which all members of the same household by definition have the same score. An example of such a variable is: *Size of the household to which you belong.* In Section 4.2.4 more information on household variables can be found. Regional variables are a very important class of identifying variables. Examples of such variables are *Residence* and *Workplace.* A category of such a regional variable defines a set of persons that may be located quite easily. Especially, when there are only a few persons in a particular region this may pose a problem; the risk of re-identification may become too high. Therefore, regional variables often play a crucial role in the SDC rules. Regional variables are examined in Sections 4.2.5 and 4.2.6. Sampling weights may provide additional identifying information to an attacker. If this is to be avoided special measures should be taken. The matter is discussed in Section 4.2.7. Remaining SDC rules are examined in Section 4.2.8. In Section 4.2.9 we give some examples of sets of SDC rules. This is done for two kinds of microdata sets: one kind that is meant for widespread use (public use files) and another kind that is meant for a select group of researchers only (microdata for research).

In Section 4.3 we examine some aspects of SDC for microdata sets in practice. These aspects include the SDC measures one could take when a certain combination of values occurs only rarely. Measures that are advocated by Statistics Netherlands are local suppression and global recoding. Application of these techniques is described in Section 4.3.1. Whenever a microdata set is released it should be avoided that the records from this set can be matched with records from other, external files. This can be done in several ways. A number of these ways are described in Section 4.3.2. When protecting a microdata set against disclosure it is often necessary to estimate the frequencies of combinations of values in the population. Because the accuracy of these estimates will generally be better for large data sets than for small ones it is advisable to use a large data set. This remark summarizes the essence of Section 4.3.3. Another issue that one encounters in practice is the protection of continuous, or quantitative, variables. When continuous variables are protected in the same way as categorical data, many combinations will be unique and therefore many SDC measures are required to protect them. In the case of categorical data we propose to check whether certain combinations of values occur frequently enough in the population. In case a continuous variable occurs in such a combination, this combination is likely to be unique in the population. However, it is not necessary to use the exact value of a continuous variable when performing the SDC checks, because it seems plausible that an attacker only knows approximate values. In Section 4.3.4 we examine the protection of continuous variables in more detail. As households are more likely to be unique in the population than the individual persons which form the households the re-grouping of household should be prevented. In Section 4.3.5 a way to impede this re-grouping is described.

4.2 SDC rules

In this section we discuss several ingredients for possible SDC rules for microdata sets in a general way. We do not specify these rules in detail, in the sense that the values of certain parameter values are specified. The reason for this is that we consider this a matter of SDC policy which is not discussed in this chapter. It should be borne in mind that our conviction is that the safety of a data set in general, and of a microdata set in particular is a gradual thing, and not something black or white. Where one should draw the line to distinguish safe from unsafe is impossible to say. It depends very much on the circumstances under which the data are being released or given access to.

As an example of the way things are discussed in this section is the suggestion to examine certain combinations of values of identifying variables, without exactly specifying which combinations are meant. This is up to the data disseminator. Likewise, we suggest to use threshold values to decide whether a combination of values of identifying variables is safe, but we do not prescribe which values to take. Again this is up to the data disseminator to define. Only in the last subsection of the present section we provide two concrete examples of sets of SDC rules. They merely serve as illustrations.

4.2.1 Classification of Variables

Variables contained in a microdata set can be subdivided into two kinds, namely variables of which the value may, possibly in combination with values of other variables, lead to re-identification, the so-called *identifying variables*, and the *non-identifying variables*. The identifying variables, or *identifiers*, themselves can in turn be subdivided into two kinds: the *direct identifiers* and the *indirect identifiers*. Examples of direct identifiers are: *Name, Address, Telephone number* and *Social security number*. A direct identifier can lead immediately to an absolutely certain re-identification when an attacker would know its value. Therefore, direct identifiers should never be published. Publication of the value of a direct identifier would make life very easy for an attacker.

On the basis of a single indirect identifier one would generally not be able to re-identify an individual. Only when the values of several indirect identifiers are used in combination this might be possible. Examples of indirect identifiers are: *Age*, *Place of residence*, *Education* and *Occupation*. A rare combination of values of indirect identifiers in a record, such as *Age=80*, *Residence=Amsterdam* and *Occupation=Farmer*, is more likely to lead to re-identification than a common one. So, indirect identifiers may be published in a microdata set, under the provision that rare combinations of values of indirect identifiers in a record are avoided. In other words, the aim of SDC for microdata sets should be to avoid the publication of certain rare combinations of values of indirect identifiers. In the remainder of this book we will mean an indirect identifier whenever we refer to an identifier, unless stated otherwise.

The particular combinations that need to be checked depend on the risk the statistical office is willing to take for the microdata set under consideration. These combinations should be described by the SDC rules for microdata sets. In practice subdividing the indirect identifiers into several classes may considerably simplify the description of the combinations that need to be checked. On the basis of such a subdivision one can systematically describe which variable combinations have to be checked. For instance, one might distinguish between three classes of identifying variables, say *identifying*, *very identifying* and *extremely identifying* ones. A variable that is *extremely identifying* is, by definition, also *very identifying*, and a *very identifying* variable is, by definition, *identifying* as well. Now one could demand that only the combinations of variables of the type *extremely identifying* × *very identifying* × *identifying* should be checked. In this way the combinations that need to be checked can be specified quite easily. In the above example it remains to be specified whether a variable is considered *extremely identifying*, *very identifying*, *identifying* or *non-identifying*, of course. In general one should describe how the variables are to be subdivided into the various classes.

Several criteria can be used to subdivide the variables into different classes. Three of such criteria are the *rareness*, the *visibility* and the *traceability* of some of the values of a variable. When the value of a variable occurs rarely in the population, then persons who score on this value are likely to become unique in the population when the values of a few other variables are also taken into consideration. An example of a variable with values that occur rarely in the population is *Nationality*. In the Netherlands there are, for instance, only very few persons with the Cuban nationality. Of course, the Cuban nationality is not rare in Cuba itself. The value of a variable is called visible when it is known, or can be ascertained easily, which persons score on this particular value. For example, the value of the variable *Sex* is visible (for most persons). The visibility of the value of a variable clearly influences the risk of re-identification. The more visible the values of a variable are, the higher the risk of re-identification. The re-identification risk is also influenced by the traceability of the values of a variable. When the persons who score on a particular value of a variable cannot be traced, then there is no risk of re-identification. In practice, however, persons that score on a particular value of a variable can, in principle, always be traced in one way or the other. What counts is the amount of work that an attacker has to perform in order to trace the persons who score on a particular value of a variable. This amount of work influences the re-identification risk. When a lot of work has to be performed to trace the persons corresponding to a value, then the associated risk of re-identification is relatively low. An example of a very traceable variable is *Residence*. It is clear, for instance, that the only place where an attacker should look for the persons who score on *Residence=Rotterdam* is Rotterdam itself.

Based on the above criteria one might consider *Residence* to be an *extremely identifying variable*, because the values of *Residence* are both very visible and very traceable. An example of a variable that one might consider to be *very identifying* is *Sex*, because its value is very visible, but not very rare and neither traceable. An example of a variable that one might consider to be *identifying* is *Occupation*, because some values are visible (*e.g.* policeman) and some other values are rare (*e.g.* professional actor) or traceable (*e.g.* public notaries of which a public file exists), but most values have neither of these characteristics.

To provide a complete set of criteria for subdividing the variables into classes seems to be impossible. Apart from some criteria one has to rely on common sense and experience gained from past surveys. For instance, although one would be tempted to consider a certain variable to be identifying, experience gained from past surveys may indicate that this variable hardly influences the risk of re-identification. In these cases it would be unwise to consider it to be identifying because this would lead to superfluous work. As an example we mention the variable *Occupation ten years ago*.

4.2.2 Rareness Criteria

As we have noted before, certain rare combinations of values of identifying variables should not be released. In this section we examine how the corresponding SDC rules should look like. These SDC rules should, for instance, prescribe the combinations that must be checked as well as the threshold values to determine whether such a combination is rare or not.

First of all, the SDC rules should prescribe the combinations that have to be checked. Preferably, the SDC rules should indicate a procedure to list all the combinations that need to be checked. In Section 4.2.1 such a procedure was sketched. The procedure in Section 4.2.1 consists of two steps. In the first step the variables are subdivided into *extremely identifying variables, very identifying variables, identifying variables* and *non-identifying variables*. In the second step the combinations that need to be checked are generated. These combinations are the crossings of an *extremely identifying variable* by a *very identifying variable* by an *identifying variable*.

Apart from describing the combinations that need to be checked, the SDC rules should also describe the threshold values to determine whether or not a combination of values should be considered rare. Here one has two possibilities: either the threshold values refer to sample frequencies or they refer to population frequencies. The threshold values should refer to sample frequencies when it is likely that an attacker knows who participated in the survey. In this case uniqueness in the sample (with respect to a certain combination of values) is sufficient to re-identify a respondent. Therefore, rare combinations in the sample should be avoided. So, a combination should be considered rare when its sample frequency is less than a threshold value S, otherwise it is considered common.

When it is not likely that an attacker knows who participated in the survey, then it seems best to let the threshold values refer to population frequencies. In this case uniqueness in the sample is not sufficient to re-identify a respondent. To re-identify a respondent this respondent should be unique in the population. So, combinations that are rare in the sample are allowed as long as they are not rare in the population. The rareness criterion should therefore, in principle, be independent of the size of the survey. For that reason, one should demand that when the population frequency of a combination that needs to be checked is less than a threshold value P this combination should be considered rare, otherwise it should be considered common.

As the threshold value P refers to population frequencies, one is usually faced with a practical problem. The population frequency of a combination may not be known. However, one has a sample available (often provided by the data set to be protected itself) to estimate the frequency of a combination in the population. Various estimators can be used to estimate the population frequency of a particular combination. In Section 5.2 some estimation methods are examined.

4.2.3 Sensitive Variables

As a working hypothesis it is not unreasonable to consider all variables in a microdata set as confidential. Therefore, the statistical office that releases the microdata set should prevent disclosure of information of individuals. Some variables should, however, be protected more severely than others. These variables are called the *sensitive variables*. Examples are variables on sexual behavior, variables on criminal past and variables on physical or mental health. A variable is called sensitive when at least one of its possible values is sensitive. Note that a variable may be sensitive as well as identifying. In the Netherlands an example of a variable that is both sensitive and identifying is *Income*. On the one hand the value of income is considered sensitive by most persons, on the other hand income can often be estimated rather accurately.

In some cases, depending on the particular microdata set and sensitive variable, it is wise not to publish this sensitive variable at all. For example, for a public use file the inclusion of a variable on the criminal past of a respondent is not recommended. In other cases, for instance when the users of the microdata set have to sign a contract declaring that they will not misuse the data to disclose information of individual respondents, some values of some sensitive variables may be published.

4.2.4 Household Variables

Variables that deserve special attention are the household variables. A household variable is a variable of which the score is equal by definition for all members of the same household. These household variables are important in case the microdata set contains records of several members of a household. When household variables are contained in such a microdata set, an attacker may use these variables to group the records that belong to the same household (assuming that the information about individuals is contained in separate records). Subsequently, the attacker can try to identify these households. As households are more likely to be unique with respect to a certain combination of values of variables than individual persons, this presents a serious threat to the privacy of the respondents. The re-grouping of households should therefore be prevented.

We can distinguish between (at least) two kinds of household variables:

1. Variables that refer to the household itself, such as the number of persons in the household, the number of under aged children and the place of residence.

2. Variables that refer to a particular person in the household, for example to the *head of the household*, and that are listed for each member of the household. Examples of these variables are the occupation and the education of the head of the household.

Examples of variables that at first sight seem to household variables but on second thought are not are *Religion* and *Political preference*. Although these variables often have the same value for members of the same household, this is not necessarily so. In any case the scores on these variables are not by definition the same for all household members. Of course they may be the same in some households.

Household variables can be identifying as well as non-identifying, for both kinds of variables may be used to re-group the records from the same household. Obviously this re-grouping only comes into play in case the microdata set contains records from at least two individuals belonging to the same household.

One way to prevent the re-grouping of households is to require that for each household of which at least two persons participated in the survey there is a sufficiently high number of other households with the same score on the household variables. Some more remarks on this method can be found in Section 4.3.5.

4.2.5   Regional Variables

Variables of which the values correspond to regions are called *regional variables*. Examples are *Residence* and *Workplace*. Regional variables are often extremely identifying, because a value of such a variable may define a set of persons that can be located easily. Therefore, the knowledge of the value of a regional variable can be very useful for an attacker.

There are in fact two kinds of regional variables, namely direct ones and indirect ones. *Direct regional variables* refer to a particular area directly, while *indirect regional variables* refer to a characteristic of an area. An example of a direct regional variable is *Residence* and an example of an indirect regional variable is *Degree of urbanization*.

Regional variables can refer to several types of regional information. For instance, one type of regional variables are variables that refer to the place of residence of a respondent. Another kind of regional variables are variables that refer to the municipality where the respondent is employed. The crossing of all regional variables, direct as well as the indirect ones, that refer to the same kind of regional information is again a regional variable. This composite regional variable offers the most detailed regional information available from the microdata set referring to that particular kind of region. Therefore, one should use this composite regional variable to perform the necessary SDC checks. This applies in particular to the indirect regional variables possibly in combination with a direct regional variable. In some cases an indirect regional variable can be very identifying when used in combination with a direct one. Suppose, for instance, that there is only one municipality X with a certain degree of urbanization in a particular province. When both the province in which the respondent lives and the degree of urbanization of the residence of the respondent would be released, then the place of residence can be determined for the respondents who live in X. Similar situations may occur when several indirect regional variables pertaining to a particular region appear in a microdata file, such as *Degree of urbanization* of a municipality, *The number of churches* in a municipality, the *Population density* in a municipality, *et cetera*. Note that if an indirect regional variable is continuous (or has very many categories), it is very likely that it in fact will act as a direct regional variable: each value is bound to be unique. Although this is coincidental and not planned, it is a fact to be reckoned with, particularly if the regional variable pertains to something that is publicly known. As an example of such a variable may serve the variable *The number of inhabitants* for a region such as municipality (at a particular reference date).

For some microdata sets, for instance public use files, one may forbid the publication of direct regional variables, because these regional variables are too identifying. In that case one should allow only some indirect regional variables. Moreover, one should limit the number of indirect regional variables: many indirect regional variables are bound to define unique areas, as we have just noted. Finally, when one decides to publish only indirect regional variables one should make sure that the regions that can be distinguished are sufficiently scattered over the country. Consider, for example, the extreme case that an indirect regional variable has only two possible values and that all municipalities in the Netherlands except The Hague assume the same value and The Hague the other one. In this case the respondents from The Hague can be recognized immediately. In the next subsection we examine how sufficient scattering of the values of indirect regional variables can be defined.

To control the risk of re-identification one could restrict oneself to the publication of only one kind of regional information. For example, one could decide to release only variables containing information on the place of residence of the respondent and no variables containing information on the municipality where the respondent is employed. Publishing different kinds of regional variables in the same file (*e.g. Residence* and *Workplace*) may lead to unacceptably high re-identification risks. The reason for this is that combinations of regional variables (not necessarily pertaining to the same kind of region) define a new compound variable in which the number of categories is the product of the number of categories of each of the regional variables in the compound; and this number can easily be quite high, so that there is a fair chance that there will be a large number of uniques.

4.2.6   The Spread of Regional Characteristics

When one adopts SDC rules that prescribe that only indirect regional variables may be published in a certain microdata set because direct regional variables are considered too identifying, then one should ensure that the values of the indirect regional variables are sufficiently scattered over the country. In other words, one should avoid that small, compact areas can be distinguished.

First of all one should decide about the smallest regions that the values of the indirect regional variables may refer to. These regions will be called the *elementary regions*. For instance, one could decide to take the municipalities as elementary regions. In this case the value of an indirect regional variable may be a characteristic of a municipality, but not a characteristic of only a part of this municipality. We assume, for the moment, that the elementary regions corresponding to a record may not be recognized, and therefore are not part of an imagined microdata file to be released. Indirect regional variables with values that refer to only a part of the elementary regions (say neighborhoods in municipalities) may not be published. In the sequel to this subsection we assume that all the indirect regional variables refer to the elementary regions or to areas consisting of a number of these elementary regions. Moreover, we will assume, for the sake of convenience, that the elementary regions are municipalities.

The crossing of all indirect regional variables of the same kind, *e.g.* referring only to the place of residence of the respondent and not to the workplace of the respondent, is again an indirect regional variable. The values of this derived variable subdivide the country in certain areas. Each of these areas consists of a number of municipalities. These areas should not be too small, *i.e.* these areas must have sufficiently many inhabitants belonging to the target population. For instance, one could demand that the number of persons in the target population in each area should be at least 100,000.

Such an area should not consist of only a few municipalities, for otherwise the re-identification risk for respondents from those municipalities could be too high. Moreover, the number of persons in the target population in such an area should not be dominated by the number of persons in the target population in only a few municipalities in that area, for otherwise the re-identification risk for respondents from the dominating municipalities would be too high. For instance, one could demand that the number of persons in the target population in any M municipalities in an area is less than $p\%$ of the total number in the target population in this area, where p is a value between 0 and 100. A possible value for M is 2 and for p 50, *i.e.* the number of persons in the target population of any two municipalities in an area should be less than 50% of the total number of persons in the target population in that area. This rule is analogous to the so-called dominance rule, which will be discussed in Section 6.2 in the context of tabular disclosure control.

Finally, the areas should consist of municipalities that are sufficiently scattered over the population. When an area would consist of a number of neighboring municipalities an attacker would have a rather compact part of the country to search for respondents. For example, the Netherlands is subdivided into twelve provinces. In this case one could demand that each area should consist of municipalities from at least four different provinces.

4.2.7 Sampling Weights

When a statistical office releases a microdata set sampling weights are sometimes included as a service to the public. A description of the auxiliary variables used, their categories and the sampling method is in that case provided as well. Unfortunately, the sampling weights, innocent as they may seem, can provide additional identifying information to an attacker. For example, when *Age* is used as an auxiliary variable and persons older than 65 years are oversampled, then the fact that a person is older than 65 years can be concluded from the corresponding sampling weight. This sampling weight will be relatively small.

When sampling weights can provide additional identifying information to an attacker, SDC measures should be taken. A rather crude way to avoid the derivation of additional identifying information is not to publish sampling weights at all. This is the safest option, of course, but one that may not be acceptable, certainly not when some subpopulations of the target population have been oversampled. Two other possibilities to avoid the derivation of additional identifying information are subsampling and the adding of noise to the sampling weights. Subsampling those records with a relatively low weight leads to an increase of these weights when re-calculated. By making all the weights approximately equal in this way they do not provide any information anymore and could as well be deleted from the microdata set. When noise is added to the sampling weights one releases a perturbed value $W_i' = W_i + \varepsilon_i$ instead of the true sampling weight W_i of record i, where ε_i is a (small) random value. As a consequence, records corresponding to the same stratum will have, as a rule, different sampling weights. The derivation of additional identifying information is therefore made much more difficult. More details on the SDC problems caused by sampling weights are presented in Section 5.5.

4.2.8   Remaining SDC Rules

In this section a number of remaining possible SDC rules is discussed. A rather primitive, but effective, way to control the risk of re-identification is to limit the number of identifying variables. For instance, one could demand that a particular kind of microdata set may contain at most 10 identifying variables. For another kind of microdata set, where one is willing to accept a somewhat higher risk of re-identification, one could demand that it contains at most 25 identifying variables.

In order to reduce the probability that an attacker knows who participated in the survey one could decide to release only microdata sets that are sufficiently outdated. For instance, one could decide to release only microdata sets that are at least two years old. Releasing only microdata sets that are sufficiently old has the additional advantage — as far as SDC is concerned — that some of the values of the identifying variables, *e.g.* the value of *Occupation*, may have been altered. This makes it more difficult for an attacker to match the records to an external file. So, releasing only sufficiently outdated microdata sets reduces the risk of re-identification.

Because the order of the records in a microdata set may reveal some additional identifying information, this microdata set should be scrambled, *i.e.* the order of the records should be randomly permuted, to reduce disclosure risk. Otherwise the order of the records in a microdata set may provide additional identifying information when, for instance, all records corresponding to the same type of region, or all records pertaining to the same household are listed consecutively. In these cases additional identifying information might be obtained on regions or households, respectively.

A special problem is posed by microdata sets obtained by panel surveys. When each microdata set is protected separately against disclosure an attacker might still be able to re-identify respondents by matching the individual microdata sets. By matching these individual microdata sets much more identifying information on respondents is obtained. For instance, when the place of residence of a respondent is contained in the microdata sets, then the respondents who have moved from one residence to another can often be recognized relatively easily. As there are usually relatively few persons that have moved from one municipality to another this may be very identifying information, not in the least since it is rather visible.

So, in fact one should protect the (imaginary) large microdata set that is obtained by matching all the individual microdata sets against disclosure. This large microdata set contains all the identifying information that is available. In the above example it contains, for instance, information on the residence of a respondent at different moments in time. When this large microdata set is protected against disclosure the individual microdata sets are automatically also protected against disclosure. Unfortunately, protecting this large microdata set is bound to lead to a lot of information loss as a result of local suppressions, global recodings and possibly other SDC measures. In particular the information loss due to such measures tends to cumulate. Suppose for instance that at a certain point in time *Residence* has to be suppressed in a particular record, as a result of a move of a household. Then in the subsequent releases of the panel sets *Residence* still has to be suppressed in the corresponding record. This is to prevent that the first suppression, applied to hide a move of the household in question, can be uncovered.

Therefore, if one decides to release microdata sets obtained through a panel survey one should resort to a more practical approach. For instance, one could decide not to incorporate any regional information in the microdata sets. This would eliminate a major source of information loss in other variables as a result of SDC measures and still keep the risk of re-identification acceptably low.

4.2.9   Examples of Sets of SDC Rules

In this section we give some examples of possible sets of SDC rules a releaser of microdata may adopt for a particular kind of microdata sets. We give two examples. The first example is a set of SDC rules that might be used for microdata sets that are released to trustworthy researchers only who have to sign a contract stating that they will not misuse such a set to disclose information about individual respondents. These microdata sets, which we will call *microdata sets for research*, may contain rather detailed information. The second example is a set of SDC rules that may be used for microdata sets that are meant for widespread use. These microdata sets, *public use files*, should not contain too detailed information.

For both types of microdata sets the SDC rules should describe which variables have to be considered identifying. One could take the criteria from Section 4.2.1, *i.e.* rareness, visibility and traceability, as a point of departure for this part of a set of SDC rules. In our example we subdivide the identifying variables for the microdata sets for research into three classes, namely *extremely identifying variables, very identifying variables* and *identifying variables*. Again, this subdivision can be based upon the criteria from Section 4.2.1. Finally, for the public use files the SDC rules should describe which variables are considered sensitive. We shall not dwell on these aspects of the SDC rules any longer and proceed with the discussion on specific examples of sets of SDC rules.

For the microdata sets for research one could use the following set of rules:

A.1. Formal identifiers should not be published.

A.2. The indirect identifiers are subdivided into *extremely identifying variables, very identifying variables* and *identifying variables* (see Section 4.2.1). Only direct regional variables are considered to be *extremely identifying*. Each combination of values of an *extremely identifying variable, very identifying variable* and an *identifying variable* should occur at least 500 times in the population. For estimating the frequency of such a combination the interval estimator in Section 5.2 could be applied.

A.3. A region that can be distinguished in the microdata set should contain at least 25,000 inhabitants.

For public use files use one could use the following set of rules. The specific form of these rules (see rule B.5) derives from the situation in the Netherlands. With suitable adaptations they should be applicable elsewhere as well.

B.1. Direct identifiers should not be published.

B.2. A microdata set must be at least one year old before it may be released.

B.3. The number of identifying variables is at most 15.

B.4. Direct regional variables should not be released. Only one kind of indirect regional variables, *e.g.* only those that refer to the place of residence of a respondent, may be published. A value of such an indirect regional variable may refer to a municipality, but not to a part of a municipality.

B.5. The combinations of values of the indirect regional variables should be sufficiently scattered, *i.e.* each area that can be distinguished should contain at least 100,000 persons in the target population and, moreover, should consist of municipalities from at least four different provinces (the Netherlands is subdivided into twelve different provinces). The number of inhabitants of any two municipalities in an area that can be distinguished should be less than 50% of the total number of inhabitants in that area.

B.6. Sensitive variables may not be published.

B.7. At least 250,000 persons in the population should score on each value of an identifying variable. At least 2,000 persons in the population should score on each value of the crossing of two identifying variables.

B.8. For each household from which more than one person participated in the survey we demand that the total number of households that correspond to this particular combination of values of household variables is at least ten.

B.9. It should be impossible to derive additional identifying information from the sampling weights.

B.10. The records of the microdata set should be published in random order.

According to these sets of rules the public use files are protected much more severely than the microdata for research. Note that for the microdata for research it is necessary to check certain trivariate combinations of values of identifying variables and for the public use files it is sufficient to check bivariate combinations. However, for public use files it is not allowed to release direct regional variables. When no direct regional variable is released in a microdata set for research, then only *some* bivariate combinations of values of identifying variables should be checked according to the SDC rules. For the corresponding public use files *all* the bivariate combinations of values of identifying variables should be checked.

4.3 SDC for Microdata Sets in Practice

4.3.1 SDC Measures

When a combination of values of variables occurs only rarely in the population, SDC measures should be taken to protect the units with this combination of values. In this subsection we consider two of these measures, namely local suppression and global recoding.

When a rare combination is encountered in a microdata set, for example the combination *Residence = Amsterdam, Sex = Female* and *Occupation = Farmer*, then SDC measures have to be taken. When local suppression is used at least one of the three values should be suppressed. For instance, one could suppress the value of *Occupation, i.e. Farmer* could be replaced by *missing*. Likewise, one could suppress the value of *Residence* or *Sex*.

When the value of a variable in a rare combination is suppressed, it remains to be checked whether the resulting combination of values is safe or not. For instance, when we suppress the value of *Occupation* in the above example it remains to be checked whether the combination *Residence = Amsterdam* and *Sex = Female* occurs frequently enough in the population, which, of course, is the case.

Global recoding is another SDC technique for protecting unsafe combinations. When a rare combination is encountered in the microdata set, then some values of one or more variables in this combination are recoded into new categories for the entire microdata set. For instance, in the above example one could recode *Occupation = Farmer* into *Occupation = Farmer or Fisherman*. Again one should check whether the resulting combination of values, in our case *Residence = Amsterdam, Sex = Female* and *Occupation = Farmer or Fisherman*, occurs frequently enough in the population. If this is not the case, then additional SDC measures should be taken.

Note that local suppression affects only some particular records, namely those records in which an unsafe combination occurs, while global recoding affects all records in which at least one of the recoded values occurs, irrespective of its combination with scores of other variables.

Another method that can be applied as a means to protect a microdata set against disclosure is the calculation of a new variable from one or more variables in the original data set. Instead of releasing all the original variables the new one is subsequently released, and possibly only some of the original ones. For instance, suppose that the original microdata set contains the variables *Residence* and *Workplace*. Maintaining both variables in the microdata set will lead to a rather high risk of re-identification. For a commuter study it could be likely that most users are only interested in the place of residence of respondents and the distance to the workplace of the respondent. Instead of publishing *Residence* and *Workplace* one could decide to publish only the distance between both municipalities, together with *Residence*. In this way the re-identification risk will be reduced considerably.

4.3.2 Matching Microdata Sets

When a relatively large microdata set is available for release, it is often possible to publish several smaller microdata sets from this base set. Each of these small microdata sets could focus on a particular part of the survey, *i.e.* each microdata set contains only some related variables. However, when several microdata sets from the same base set are released, it may be possible that the variables in common can be used to match these sets. The re-identification risk would be increased considerably in that case. Therefore, matching several microdata sets obtained from the same base set should be impossible, or at least very hard to realize, for an attacker. Note that for matching several microdata sets an attacker can use both the identifying variables and the non-identifying ones.

There are several ways to prevent the matching of microdata sets obtained from the same base set. First of all, one can protect the base set against disclosure. In this case each microdata set, or combination of microdata sets, obtained from the base set is also protected against disclosure. In other words, even when several microdata sets are matched the resulting microdata set is safe. A disadvantage of this approach is that too many SDC measures may be necessary to protect the base set.

A method to prevent the matching of microdata sets obtained from the same base set is to release only records of disjoint sets of persons. A record of a particular person occurs in at most one of the released microdata sets. In this case it is clearly not possible to match the released microdata sets. A disadvantage of this method is that the total number of released records amounts to at most the number of records in the base set. So, each released microdata set will contain (considerably) less records than the base set.

The matching of microdata sets obtained from the same base set can also be prevented by releasing only disjoint sets of variables. A particular variable occurs in at most one of the released microdata sets. Again it is clear that it is not possible to match the released microdata sets. A disadvantage of this method is that the total number of released variables amounts to at most the number of variables in the base set. So, each released microdata set will contain (considerably) less variables than the base set. Note that we do not only refer to the identifying variables but also to the non-identifying ones. Non-identifying variables can also be used to match microdata sets.

Finally, we can check whether each combination of values of variables occurs in more than one record in a released microdata set. Exact matching is not possible in this case, because the records that correspond to a particular person cannot be determined with certainty. The combination of values of both the identifying and the non-identifying variables should be examined in this case, because non-identifying variables can be used to match microdata sets just as well as the identifying ones.

4.3.3   The Use of a Larger Microdata Set

Because it is usually necessary to make estimates while protecting a micro-data set against disclosure, it is advisable to use a large data set to base these estimates upon. The estimates based on large data sets will be more reliable than those based on small ones. As a result the SDC measures will be more appropriate. Therefore, if one wants to protect a relatively small data set one should look for a larger data set that contains (part of) the same variables and base (some of) the estimates on this larger set. Note that it is not necessary that the larger data set contains the same records, or records of the same persons, as the small one. It is only necessary that (part of) the variables are contained in the smaller data set as well as in the larger one.

One should be careful when using a larger data microdata set to protect a relatively small one. In that case one should make sure that both data sets refer to the same target population. For example, when the target population of the small data set consists of those persons in the population whose occupation is teacher and one has a larger microdata set containing also information on other occupations, then one should select only those records from the large microdata set that refer to persons whose occupation is teacher.

4.3.4   Continuous Variables

Continuous, or quantitative, identifying variables can be protected against disclosure in a similar way as categorical variables. In fact, continuous variables are also categorical, or qualitative, variables because for a finite population they assume only a finite number of values. Each value can be considered to be the category of a categorical variable. However, treating continuous identifying variables in the same way as categorical identifying variables would generally lead to many modifications in the microdata set. Moreover, the microdata set would be protected too severely in most cases. For a continuous identifying variable we may assume that the exact score of a respondent on this variable is not exactly known by an attacker, but only approximately. Therefore, it is not necessary to use the exact value of a continuous identifying variable when protecting a microdata set. We propose to protect continuous identifying variables in two steps against disclosure. In a first step the domain of a continuous variable is partitioned into a number of classes. The way the classes are formed should reflect the approximate knowledge of an attacker with respect to the continuous variable. The new classes are the categories of a new variable. This categorical variable can then be protected against disclosure in the way that is usual for categorical variables. In a second step the extreme values of the continuous variable are protected against disclosure. The first step is examined in the remainder of this section, the second step is examined in Section 5.3.

The values of a continuous identifying variable, such as *Age* (in days or even years), are usually not known exactly to an attacker. Instead an attacker will only know approximate values. In the case of the age of a person an attacker may be able to estimate the age plus or minus two years, but he is generally not able to say the actual age. For instance, an attacker may be able to say that the age of a person lies between 30 and 34 years, but not that the actual age is 32. So, an attacker is often only able to say in which class the value of a continuous variable lies. For some variables, such as *Annual income*, the exact value is often unknown even to the respondent. After protecting these classes by means of the usual techniques for categorical variables the actual values may be published.

Let us suppose that we want to protect the variable *Age*. We assume that an attacker is able to estimate the age of a person plus or minus two years. We assume that such a *window* of possible ages is symmetric around the actual age, and that the length of the window is 5 (for persons of 0 or 1 year the lower bound of the window is chosen to be 0, and the upper bound the actual age plus 2). That is, for a person of 32 we assume that an attacker is able to determine that the age is 30, 31, 32, 33, or 34. For each person a window of possible ages can be determined by an attacker. These windows can be seen as categories of a new variable *Age-class*. Any person (older than 1) scores on five categories of *Age-class*. For example, a person of 32 scores on the categories 28–32, 29–33, 30–34, 31–35 and 32–36 of *Age-class*, and a person of 84 scores on the categories 80–84, 81–85, 82–86, 83–87 and 84–88 of *Age-class*. We say that category 30–34 of *Age-class corresponds* to category 32 of *Age*, and that category 82–86 corresponds to category 84.

Next we protect the variable *Age-class* by means of the usual techniques for categorical variables. On the basis of the protections carried out involving *Age-class* we decide what to do with *Age*. When a category of Age-class occurs sufficiently often in the population, then the corresponding value of *Age* is considered safe. For example, when the category 30–34 of *Age-class* occurs frequently enough in the population, then category 32 of *Age* is considered safe. When the category 82–86 of *Age-class* does not occur frequently enough, then category 84 of *Age* is considered unsafe.

If some categories of *Age-class* have been recoded then the corresponding values of *Age* may not be published. Instead a suitably chosen recoding of *Age* should be published. For instance, suppose that category 82–86 of *Age-class* is considered unsafe, and that the category obtained by collapsing the three categories 82–86, 83–87 and 84–88 is considered safe, *i.e.* there are sufficiently many persons with age 82, 83, 84, 85, 86, 87, or 88 in the population. A suitable recoding of the original variable *Age* is obtained by collapsing the categories corresponding to the collapsed windows. In this case, this implies that categories 84, 85 and 86 of *Age* should be collapsed. Note that when collapsing categories from *Age-class* one should make sure that each window is combined at most once with other windows, for otherwise a strange categorization of *Age* would be obtained.

If some local suppressions on *Age-class* have been carried out, then *Age* can be published, provided that in the records in which a category of *Age-class* is suppressed the corresponding *Age*-value will be suppressed.

Other continuous variables like *Annual income* can be protected in the same way. Usually problems occur in the tails of such variables, and the corresponding treatment of such problems goes under the name of top/bottom-coding. This problem is considered in more detail in Section 5.3.

4.3.5 Comments on the Construction of Household Files

When protecting a microdata set against disclosure one should start by preventing the re-identification of the individual persons by means of the identifying variables. However as indicated before, this is not sufficient, because it may sometimes be possible to group the records obtained from persons of the same household. This possibility to group records obtained from the same household increases the risk of re-identification, because households are more likely to be unique than individuals. Therefore, this possibility to match records obtained from the same household should be prevented, or at least be made more difficult. A method to do so is the following.

Start by constructing a file of *household records*. This can be done by using direct household identifiers, such as a unique household number, contained in the original, unprotected, microdata set. This household file should contain one record for each household. This record consists of the household variables and the number of persons from this household that participated in the survey. From this set of household records remove the records where only one person from the household participated in the survey, because these records are protected when the individual records from original microdata set are protected. Check for each record in the remainder of the household set whether there is a sufficient number of other records with the same score on the household variables. Here a sufficient number means more frequently than a certain threshold value H. For instance, one could take $H = 5$ or, if one wants more protection, $H = 10$. When there are records in the remainder of the household set that do not satisfy the above criterion then some values of the household variables should be suppressed, some records should be deleted from the original microdata set or some household variables should be recoded until the above criterion is satisfied. In this way it is hard to match records obtained from the same household.

After the matching of records obtained from the same household has been prevented in the above way the direct household identifiers are deleted and the modified microdata set can be published (in case the other requirements of the SDC rules are satisfied).

5

Microdata: Backgrounds

Every individual nature has its own beauty.
—Emerson

5.1 Introduction

The present chapter discusses some theoretical issues related to SDC for microdata. The discussion is topical rather than systematic. It also merely aims at bringing certain aspects involved to the attention of the reader rather than treating them in-depth. An in-depth treatment would involve rather advanced mathematics, something that this book tries to avoid as much as possible.

Although most of the issues raised here concern an individual that is charged with producing "safe" data this does not hold for all of them. A topic that can be considered as a theoretical issue is the calculation of disclosure risks for microdata sets. This is merely something that can be used to gain some theoretical insight into the factors that contribute to the risk of disclosure when a microdata set is in the hands of an attacker. The model involved also gives general clues as to how to reduce this risk, but it should be considered too crude for real practical use. A much more useful model would be one that yields an expression for the risk for each individual record in a microdata set, as was observed in Chapter 2. As such a model, that can also be used in practice, does not seem to be available at the time of writing, we should settle for less. The cruder model discussed in Section 5.6 could be considered as a first step towards a more refined model for calculating disclosure risks of individual records in a microdata file.

Although the practitioner charged with the production of a "safe" microdata set is confronted with most of the issues discussed in this chapter, it should be clear from the discussion below that dealing with them is often no trivial exercise, in particular if specialized software is lacking.

A very important aspect of SDC for microdata sets is the estimation of population frequencies of a combination of values of identifying variables. This presents a problem especially when the frequency of such a combination in a small population, e.g. in a small region, must be estimated. In Section 5.2 a number of estimators, namely the direct point estimator, a synthetic estimator, a compromise estimator and an interval estimator, are presented.

The extreme values of continuous variables may lead to re-identification of the respondent. Therefore, such extreme values should be protected especially well. A method for protecting the extremely high values, top-coding, and extremely low values, bottom-coding, of a continuous variable is examined in Section 5.3.

A general discussion on global recoding and local suppression is presented in Section 5.4. In particular it is sketched how the minimum number of necessary local suppressions may be determined. For this the concept of minimum unsafe combinations is introduced. Using this concept it is not difficult to formulate the problem of determining the minimum number of necessary local suppressions as a 0-1 integer programming problem.

In Section 5.5 it is illustrated how the sampling weights can provide additional identifying information to an attacker. This is done only for the case where the sampling weights have been determined by post-stratification. For other methods to determine sampling weights, such as raking, it is generally also possible to derive additional identifying information from the sampling weights. Two methods to prevent this derivation of additional identifying information, subsampling and adding noise, are explained.

Section 5.6 discusses a general model to evaluate the re-identification risk for an entire microdata set. This section is rather theoretical in nature. It is included in this book because several national statistical institutions apply similar models to base the SDC measures for their microdata sets upon.

5.2   Estimation of Population Frequencies

For the disclosure control of microdata it is generally required that certain combinations of characteristics, i.e. combinations of values of variables, occur frequently enough in the population. If such a combination of characteristics does not occur frequently enough in the population, the records in which these combinations occur are considered "unsafe", and appropriate SDC measures should be taken to reduce the associated disclosure risk, e.g. by global recoding of variables or local suppression of values. A problem in practice is often that it is not known what the frequency of a particular combination of values in the population is, contrary to the frequency of its occurrence in the sample, which can easily be found out. This population frequency therefore has to be estimated from a sample. This sample should ideally be a large one in order to obtain accurate estimates. But sometimes no big sample is available (pertaining to the same population) and one has to make estimates from the microdata set one wants to safeguard through an SDC treatment.

In this section we consider some methods for estimating the frequencies mentioned above. First we consider a direct point estimator. For small sample cell frequencies, however, the variance of this direct point estimator is too large for this estimator to be reliable. Therefore, compromise estimators are introduced that try to remedy this defect. Each of these compromise estimators is a convex combination of a direct estimator and a so-called synthetic estimator. The weights of these convex combinations are determined by the mean square errors of the respective component estimators. In addition to these point estimators an estimation method based on a Poisson model for the cell frequencies is examined. A feature of this estimation method is that it is an interval estimation procedure. Such an estimation method allows one to incorporate the accuracy of the estimator, something that is lacking in a point estimator. Although the accuracy can be incorporated in this interval estimator, it nevertheless has its shortcomings, namely when the sample cell frequencies are small, for instance in the case of small areas. In that case compromise estimators are usually better.

We now start, in a first attempt, to calculate a sample threshold value. Let Y_z be the number of persons in a population with a certain combination of values z, for short *with property* z. For the remainder of this section we shall assume that z is a combination that has to be checked according to some SDC rules. A requirement for such a combination z to be considered safe could be that Y_z is sufficiently large, *i.e.* by application of the

$$\text{\textbf{Rule:} } z \text{ is safe if and only if } Y_z \geq d, \tag{5.1}$$

for a certain threshold value d. The value of d can be chosen in view of the particular purpose of the microdata set to be treated. In general a criterion such as 5.1 cannot be used directly because Y_z is not known for property z. What we could try to do is to replace Y_z in 5.1 by an estimator $\hat{Y}_z$. Instead of criterion 5.1 we would have

$$\text{\textbf{Rule:} } z \text{ is safe if and only if } \hat{Y}_z \geq d, \tag{5.2}$$

The problem now is to find good estimators for Y_z. Actually, we will try to find good estimators for $Y_{i,z}$, *i.e.* the frequency of the occurrence of z in region i in the population. In this way our notation agrees with the notation used in the literature on small area estimation.

A first idea that comes to mind is to use a direct estimator. Let $y_{i,z}$ be the frequency of the occurrence of property z in region i in the sample,

and n_i the number of individuals in the sample in region i. The fraction
of persons with property z in region i, denoted by $\nu_{i,z}$, can be estimated
by $\nu_{i,z} = y_{i,z}/n_i$. An estimator of the number of persons in region i with
property z would be

$$\hat{Y}_{i,z} = \nu_{i,z} N_i, \tag{5.3}$$

where N_i is the number of persons in the population living in region i,
which is assumed to be known. This point estimator $\hat{Y}_{i,z}$, that we shall
refer to as the direct estimator, is very unreliable if the number of persons
in the sample from region i is small. In practice, regions are often very
small in this sense, and therefore we have to find estimators that are able
to cope with this situation.

If we assume that the fraction of individuals with property z is spread
more or less homogeneously over the population, we can pool the regions
and estimate the relative frequency of the occurrence of z by

$$\nu_z = \frac{\sum_i y_i}{\sum_i n_i}, \tag{5.4}$$

so that we would obtain as an estimator for Y_z

$$\hat{Y}_{i,z} = \nu_z N_i. \tag{5.5}$$

The estimator 5.5 is called a synthetic estimator; it has the advantage
that it has a smaller variance than the direct estimator. However, estimator 5.5 is biased if the assumption that property z is spread homogeneously
over the population is not satisfied.

In order to meet the observed difficulties with the direct and synthetic
estimator, it is possible to construct an estimator that is a weighted sum of
these estimators. Such an estimator is called a compromise (or combined,
or composite) estimator. An estimator $\hat{\nu}_{i,z}$ for the relative frequency of
property z in region i is then given by

$$\hat{\nu}_{i,z} = W_i \nu_{i,z} + (1 - W_i)\nu_z, \tag{5.6}$$

where $0 \leq W_i \leq 1$ is chosen in such a way that the mean square error
(MSE for short) of $\hat{\nu}_{i,z}$ is minimal, $i.e.$

$$W_i = \frac{MSE(\nu_z)}{MSE(\nu_{i,z}) + MSE(\nu_z)}. \tag{5.7}$$

The mean square error of the compromise estimator $\hat{\nu}_{i,z}$ is given by

$$MSE(\hat{\nu}_{i,z}) = W_i^2 MSE(\nu_{i,z}) + (1 - W_i)^2 MSE(\nu_z). \tag{5.8}$$

Note that the MSE of the compromise estimator depends on the MSE
of both the direct estimator and the synthetic estimator. If the sample

is assumed to be of the simple random sampling type, the MSE of the synthetic estimator depends both on its bias and on its variance, whereas the MSE of the direct estimator equals its variance since it has zero bias. If property z is reasonably homogeneous, the bias of the synthetic estimator will be small. Since this estimator has a smaller variance than the direct estimator, the contribution of the synthetic estimator in the compromise estimator will dominate that of the direct estimator. In that case we will have that W_i is close to 0. For less homogeneously distributed properties z the reverse will be the case and the contribution of the direct estimator will dominate that of the synthetic estimator, due to the (large) bias of the synthetic estimator. In that case the value of W_i will be closer to 1.

In practice, the mean square error of the direct estimator and the synthetic estimator are not known exactly. So, they have to be estimated. Often a probability model is assumed in order to do so. The expected value of the MSE of both the direct estimator and the synthetic one under the model assumption is then used as estimates for the true values of $\mathrm{MSE}(\nu_{i,z})$ and $\mathrm{MSE}(\nu_z)$. This approach is used in, for example, [17] and [57]. For a general introduction to compromise and other estimators for small domains we refer to [7].

Another useful idea is to work with interval estimators rather than with point estimators as above. The advantage of an interval estimator is that the accuracy of the corresponding point estimator is explicitly taken into account. As was remarked before, point estimators for small sample sizes have a relatively large margin. This means that there is a considerable probability that the estimated value of $Y_{i,z}$ is larger than d, while its true value is smaller than d. In that case there would be no application of an SDC measure which in fact would be necessary. Confidence regions can be determined, such that 5.1 can be replaced by a similar rule "at sample level":

$$\textbf{Rule: } z \text{ is safe if and only if } y_z \geq d_s, \qquad (5.9)$$

where d_s is a threshold value. This value is determined such that the conditional probability that the number of persons in the sample with property z is greater than d_s, given that the number of persons in the population is smaller than d, is bounded from above by a certain chosen value α.

Suppose that a property z is considered rare when its (unknown) population frequency $Y_{i,z}$ is less than the threshold value d. We test the hypothesis

$$H_0: \quad Y_{i,z} < d, \qquad (5.10)$$

against the alternative hypothesis

$$H_1: \quad Y_{i,z} \geq d, \qquad (5.11)$$

If H_0 is rejected at a certain significance level α, we assume that property z occurs frequently enough in region i. The probability that this assumption is incorrect is at most α.

We assume that the (known) sample frequency $y_{i,z}$ of the combination is a realization of a stochastic variable $\underline{y}_{i,z}$. H_0 is then rejected when

$$y_{i,z} \geq d_s, \tag{5.12}$$

where d_s satisfies

$$\Pr[\underline{y}_{i,z} \geq d_s \mid Y_{i,z} < d] = \alpha. \tag{5.13}$$

It remains to specify the distribution of $\underline{y}_{i,z}$. A natural choice for this distribution is a Poisson distribution with parameter $\mu = E(\underline{y}_{i,z})$. The most likely estimate for μ is given by $y_{i,z}$. Using this estimate the value of d_s can be evaluated. In turn this enables us to check whether H_0 should be rejected or not. In this way we can determine if a certain property should be considered rare.

5.3   Continuous Variables

Extreme values of continuous, or quantitative, variables, such as *Income* and *Age*, can lead to disclosure. Therefore, these extreme values have to be protected against disclosure. A method to safeguard extreme values is described below. Safeguarding extremely large values against disclosure is called top-coding; safeguarding extremely small values is called bottom-coding.

Suppose we have to check whether the combinations given by values of *Residence* × *Sex* × *Income* occur frequently enough in the population according to the SDC rules. The variable *Income* is, however, a continuous variable. So, taking the exact value of *Income* into consideration will lead to many unique combinations. It is better to take classes of values of *Income* into consideration while checking the combinations, as we suggested in Section 4.3.4. However, the extreme values of *Income* remain to be checked. For this the following SDC rule can be applied: a combination of values *Residence=D*, *Sex=G*, *Income=X* is considered safe when there are at least N_X persons in the population with *Residence=D*, *Sex=G* and *Income=Y* where Y is "comparable" to X. Here two values X and Y of *Income* are considered *comparable* when they differ less than $p\%$ (with respect to the highest value of X and Y). Note that a value X is comparable to itself by definition . By selecting values for N_X and p, for instance $N_X = 100$ and $p = 20$, this SDC rule can be applied in practice. When $N_X = 100$ and $p = 20$ is used, there should be at least 100 persons in the population with an income that differs less than 20% from X.

The above mentioned rule requires that certain combinations occur frequently enough in the population. However, in practice we often only have the microdata set to be released available to check this. In that case we have to estimate these frequencies. This can be done by means of a point estimator or an interval estimator (*cf.* Section 5.2). When one wants to apply an interval estimator another parameter, α, should be chosen. This parameter α equals the significance level of the test corresponding to the interval estimator. In other words, we test whether there are at least d values in the population that differ less than $p\%$ from the largest value. So, the probability that an extreme value is unjustly considered safe is less than α.

Suppose we need to protect a microdata set. The sampling fraction of this microdata set is $1/150$. The microdata set contains the three variables *Residence, Sex* and *Income*. Suppose we need to check whether the combinations given by values of *Residence* × *Sex* × *Income* occur frequently enough in the population. Moreover, suppose we use the parameters $N_X = 100$ and $p = 20$. We decide to apply an interval estimator to determine whether such a combination occurs frequently enough in the population. For this we assume that the stochastic variable *Residence* × *Sex* × *Income* is Poisson distributed. We test the hypothesis

H_0: The combination *Residence=D, Sex=G* and *Income comparable to X* occurs less than 100 times in the population.

against the alternative hypothesis

H_1: The combination *Residence=D, Sex=G* and *Income comparable to X* occurs at least 100 times in the population.

The significance level α of this test, *i.e.* the maximum probability that H_0 is rejected while it holds true, is chosen to be 0.1. This implies that H_0 is rejected for a particular combination *Residence=D, Sex=G* and *Income=X* when there are three or more combinations with *Residence=D, Sex=G* and *Income is equal or comparable to X in the microdata set*. When H_0 is accepted SDC measures must be applied, like suppressing the value of *Residence*. When H_0 is not rejected SDC measures do not have to be applied for this particular combination of values.

Stated in more formal terms the extremely high values of a continuous variable should obey the following rule.

Rule 1: There should be at least d population elements with a value in the interval $[(1-p)h, h]$ where h is the largest observed value in the sample, for some prespecified value p in the interval $(0,1)$.

Unfortunately, Rule 1 can only be applied if the population values are known. In practice we generally know the sample values only. In that case it is only possible to evaluate the probability that Rule 1 is satisfied. So, instead of Rule 1, the following rule seems to be more useful.

Rule 2: The probability that Rule 1 is incorrectly assumed to be satisfied
is at most α, where α is a small number.

Assume there are n observations x_i of a continuous variable X from a
population of size N, such that $x_1 \geq x_2 \geq \cdots \geq x_n$. Let f_s denote the
fraction of values in the *population* between x_s and x_1. Because x_1 and x_s
are random variables, namely the 1-st and s-th rank statistic respectively,
also the number of population elements between x_s and x_1 is a random
variable. Consider now the probability $(1 - \alpha')$ that for a particular value
s f_s is at least equal to the fraction $\gamma = d/N$, *i.e.*

$$\Pr[f_s \geq \gamma] = 1 - \alpha'. \tag{5.14}$$

If $s = 1$, the width of the interval is 0 and $1 - \alpha'$ is 0 as well. If s increases
so does $1 - \alpha'$. The smallest value of s such that $1 - \alpha'$ is at least equal to
$1 - \alpha$ will be denoted by k. Let I be the interval in Rule 1, *i.e.* $((1-p)x_1, x_1)$.
If $(1-p)x_1$ is smaller than or equal to x_k, the interval I contains the interval
(x_k, x_1) and the probability that I contains at least d population elements
is at least equal to $1 - \alpha$. However, if $(1 - p)x_1$ is greater than x_k, then
the interval I will contain at least d population elements with a probability
less than $1 - \alpha$. It follows that, if we find that $(1 - p)x_1$ is at most x_k,
we accept the null hypothesis that Rule 1 is satisfied. Likewise, if we find
that $(1 - p)x_1$ is larger than x_k, we reject the null hypothesis that Rule 1
is satisfied.

The critical value k for s can be found by calculating $1 - \alpha'$ for a number
of increasing values of s. The value of k depends on the given values of d, p
and α. Furthermore k depends also on N (via $\gamma = d/N$) and on the sample
size n. Kendall and Stuart (*cf.* [45], p. 548) show without any assumptions
with respect to the probability distribution of X that f_s is Beta distributed
with parameters $s - 1$ and $n - s + 2$, *i.e.*

$$\Pr[f_s \geq \gamma] = \int_{\gamma}^{1} \beta(y; s - 1, n - s + 2)dy, \tag{5.15}$$

where

$$\beta(y; s - 1, n - s + 2) = \frac{y^{s-2}(1 - y)^{n-s+1}}{\int_0^1 y^{s-2}(1 - y)^{n-s+1}dy}. \tag{5.16}$$

The probability given by 5.15 can be well approximated by a function
depending only on d, p, α and the sampling fraction $f = n/N$, *i.e.* this
approximation does not depend on the actual values of n and N but only
on the fraction n/N. We do not give this approximation. Instead we restrict
ourselves to presenting Table 5.1 which is based on this approximating
function. Table 5.1 is taken from [55] and shows the values of k for $d = 100$,
$\alpha = 0.05$, $p = 0.2$ and various sampling fractions f.

TABLE 5.1. Values of k for $d = 100$ and
$\alpha = 0.05$, for various sampling fractions f

$1/f$	k
≥ 1950	2
$1951 - 282$	3
$281 - 123$	4
$122 - 74$	5
$73 - 51$	6
$50 - 39$	7
$38 - 31$	8
$30 - 26$	9
$25 - 22$	10
$21 - 19$	11

Thusfar we have examined the protection of extremely high values, top-coding. The protection of extremely low values, bottom-coding, can be done in a similar way. In fact, by replacing the values of a variable that should be protected by means of bottom-coding by their opposite values top-coding can be applied to these negative values.

5.4 Global Recoding and Local Suppression

One of the most basic elements of the approach to SDC described in the present book is the checking of certain combinations of key variables. The SDC rules that are applied to produce a safe microdata set are supposed to prescribe what combinations have to be checked, and also what the minimum frequency of occurrence of a combination of values should be. Application of such SDC rules to a microdata set would yield lists of "rare" combinations of values, and records in which they appear. But identification of these problematic cases is not enough; something should be done about them, that is they should be eliminated from the file. There are several ways to eliminate such combinations from a microdata file. The ones we consider are global recoding and local suppression. These actions can be applied separately or in combination.

So suppose that we want to eliminate a set of rare combinations of values, appearing in various records in a given microdata set, by application of global recodings and/or local suppressions. It is clear that application of these techniques reduces the information contents of a microdata file. So, they should be applied parsimoniously. When we elaborate this idea it is natural that we arrive at formulating the problem of eliminating the rare combinations of a microdata file as an optimization problem. The problem is then, informally stated, how to apply global recodes and local suppressions in such a way that the resulting file is safe (*i.e.* does not contain any rare combinations) with a minimum information loss from the original file. We shall not formally elaborate this problem in the current book, because it is rather technical. Instead we shall discuss the problem rather informally and intuitively.

Suppose for the moment that we apply local suppressions only to eliminate a set of rare combinations from a file. The easiest way to determine which values of the variables should be locally suppressed would be to do this for each combination that has to be checked and for each record separately. This can be done in two ways. Firstly, when a value is suppressed, then this value is set to *missing* immediately. The resulting microdata set is then used to determine whether or not a combination is safe. Secondly, the original microdata set can be used to determine whether or not a combination is safe. However, either way causes problems.

When the first method is used some combinations may incorrectly appear to occur not frequently enough. For example, if we suppress the value *Baker* in the combination *Baker* × *Foreigner* then this may have the consequence that later on the combination *Baker* × *Male* appears to occur not frequently enough. This combination would therefore be considered to be unsafe. However, this combination could occur frequently enough in the original microdata set. So, in fact it should be considered safe.

On the other hand, when the second method is used to determine whether a combination is safe and one does this for each record separately then this may also lead to problems. Suppose for instance that the combination *Baker* × *Foreigner* does not occur frequently enough in the (original) file and that we decide to suppress *Foreigner*. Suppose furthermore that the combination *Baker* × *Female* neither occurs frequently enough and in this case we decide to suppress *Female*. Then it is not unlikely that we suppress too much. In case there would be persons who are *Baker*, *Female* and *Foreigner* simultaneously it would have been better if we had suppressed the value *Baker* for these persons assuming each of the values *Foreigner* and *Female* occurs frequently enough. The number of local suppressions would have been less if we had suppressed *Baker* for these persons.

We conclude that we cannot decide for each unsafe combination and record *separately* which values should be suppressed if we want to minimize the number of local suppressions. We have to decide which values have to be suppressed for all unsafe combinations and records *simultaneously*. This can be done quite easily by introducing so-called minimum unsafe combinations.

To fix our minds we suppose that it is necessary to check whether certain trivariate combinations of values of identifying variables occur frequently enough in the population, *i.e.* more frequently than a certain threshold value d. We start by checking all the univariates. In case a value of a variable is considered safe we check the bivariate combinations in which this variable occurs. In case a value of a variable does not occur frequently enough, e.g. *Mayor*, we do not have to check the bivariate combinations involving *Mayor*, e.g. *Mayor* × *Female*, because they will be unsafe as well. Then we check the trivariate combinations in which only safe bivariate combinations occur. In case an unsafe bivariate combination occurs in a trivariate combination then this trivariate combination is also unsafe, something that we can deduce and that needs not to be checked. For example, if the bivariate combination *Baker* × *Female* is unsafe, then evidently also unsafe is the trivariate combination *Baker* × *Female* × *Amsterdam*, and hence it is not necessary to check it. After checking the required trivariate combinations we are able to list for all records the unsafe univariate, bivariate and trivariate combinations. The combinations in this list will be called *the minimum unsafe combinations*, because whenever we suppress a value in a minimum unsafe n-variate combination, then the resulting $(n - 1)$-variate combination will be safe. This property of the minimum unsafe combinations can be used to formulate the problem of minimizing the number of local suppressions as a 0-1 integer programming problem quite easily. Namely, the target function to be minimized is given by the number of local suppressions and the constraints are given by the requirement to suppress at least one value for each minimum unsafe combination. Likewise, the problem of minimizing the number of different categories that are suppressed can be formulated as a 0-1 integer programming problem by using the concept of minimum unsafe combinations. In this case the target function is given by the number of different categories.

If the number of unsafe combinations for a microdata set is large, then it is recommended that at least some (if not all) of these combinations are removed by global recodes. If all original unsafe combinations would be removed by local suppressions, then too many missing values would have been introduced into the data set. If, after application of these recodes, there are still some rare combinations left in the microdata set, they can be removed through the application of suitable local suppressions. This latter step is best carried out automatically to avoid the problems mentioned above.

One could also try to replace the "manual" global recoding step by an automated procedure. In that case one has to develop a rather complicated optimization model first which aims to find the optimal mix of global recodes and local suppressions. In order to do this some provisions have to be made. First of all, one should introduce a suitable information measure to quantify the information loss due to a global recoding or due to a local suppression. In fact, entropy can be used for this purpose. This allows one to quantify the trade-off in information loss when applying recodings and local suppressions. In principle one can find the optimal mix of global recodings and local suppressions for a given set of rare combinations by first considering every possible partitioning of this set into two sets, by solving "pure" recoding and "pure" suppression problems for each of these sets (two possibilities for each partitioning), by quantifying the information loss for each possibility, and by finally selecting the partitioning plus optimal procedures for each partitioning set that gives a minimum information loss. In practice, however, this may be impossible because of the enormous amount of possible partitionings for a given set (2^n if the set has n elements). So one has to resort to heuristics to cope with this problem, and settle for less than an optimal solution. For details the reader is referred to [28].

It should be noted that there is a considerable similarity between the theory of optimal local suppressions and that of data editing (*cf.* [27]. In data editing the aim is to de-activate all edits that are violated by a set of records, through setting a minimum number of values in these records to missing (*cf.* [77]).

5.5   Sampling Weights

There are several methods to compute sampling weights: post-stratification, linear weighting and multiplicative weighting (the latter also called raking, raking ratio estimation or iterative proportional fitting). Details on linear weighting can be found in [4] and on multiplicative weighting in [30]. In the sequel to this section we will only show how an attacker may proceed in order to derive additional identifying information from the sampling weights in case of post-stratification when the attacker has a sufficiently precise knowledge of the population frequencies of the strata of the auxiliary variables. In case of linear or multiplicative weighting an attacker is also able to obtain additional identifying information from the sampling weights. However, these cases are somewhat more difficult and are not explained in this book. An extensive discussion of the problem addressed to in this section can be found in [26], from which the following example is also taken.

Suppose that two auxiliary variables A and B have been used to calculate the sampling weights by means of poststratification. The number of categories of A is two and of B it is three. The population stratum sizes of each of the six strata are given in Table 5.2. The weights are listed in ascending order in Table 5.3.

The weight of a stratum multiplied by its corresponding frequency in the sample is by definition equal to the size of this stratum in the population. So, when the intruder would know the population stratum sizes as given in Table 5.2, then he would be able to determine which weight corresponds to which stratum. For instance, it is easy to see that weight 82.095 corresponds to stratum $A_2 \times B_3$, and weight 89.596 to stratum $A_1 \times B_3$.

When the intruder would know the frequencies of the strata in the population only approximately, then he would have to choose the most likely way to match the weights with the strata. If the knowledge of the intruder

TABLE 5.2. Population stratum sizes

Stratum	Size
$A_1 \times B_1$	1,368
$A_1 \times B_2$	725
$A_1 \times B_3$	896
$A_2 \times B_1$	2,633
$A_2 \times B_2$	2,787
$A_2 \times B_3$	1,642

TABLE 5.3. Weights of the strata

Stratum no.	Frequency (sample)	Weight	Weight × Frequency (rounded)
1	20	82.095	1,642
2	10	89.596	896
3	29	96.102	2,787
4	25	105.320	2,633
5	6	120.833	725
6	10	136.799	1,368

about the frequencies of the strata in the population is sufficiently precise, then he will be able to determine which stratum belongs to a specific weight. Suppose, for instance, that the intruder would think that stratum i occurs $X_i = X_{pop,i} + \varepsilon_i$ times in the population, where $X_{pop,i}$ is the actual frequency of stratum i in the population and ε_i is an error term with $-50 \le \varepsilon_i \le 50$. It is easy to see that in this case the intruder would still be able to match the weights to their corresponding strata correctly.

In the case of linear weighting or multiplicative weighting an attacker can also match the sampling weights to the strata of the auxiliary weights in many cases. This is, however, more complicated than in the case of post-stratification.

When the sampling weights can lead to additional identifying information, then SDC measures have to be taken. In particular the sampling weights should be adapted. For this there are two techniques, namely subsampling and adding noise. By subsampling the records with a low sampling weight and by then re-calculating them, these weights are increased. In this way one can make all the weights approximately equal and then discard them. One can also make all the weights of the records almost equal in a first step and then apply the second method, namely adding noise to the (adapted) weights. Subsampling leads to a loss of information, of course, because several records will not be published.

The second method for reducing the risk of disclosure caused by sampling weights is adding noise to these weights. In other words, instead of releasing the true sampling weight W_i of a record i one releases a perturbed value $W_i' = W_i + \varepsilon_i$, where ε_i is a (small) random value. Adding noise leads to a loss of information, because only perturbed sampling weights are published and not the true weights.

Of course, the information that can be obtained by an attacker who has been successfully using sampling weights to identify to which strata the respondents in a microdata file correspond, does generally not lead to re-identification directly. However, it could be a step in an effort that might lead to a successful disclosure, by carefully piecing together all bits of information that can be distilled from the data and from other sources. Because sampling weights are not too difficult to protect by the SDC measures indicated, there is no fundamental reason in not doing so. If the threat to privacy posed by non-protected sampling weights is considered too futile a problem, then consider the embarrassment that can be prevented to the data releaser. It looks a bit silly if somebody can easily deduce certain information that was not supposed to be in a file. For instance a public use microdata file that is sold as a file without any information pertaining to the place of residence of the respondents, because it does not contain any such regional variables, should neither contain such information implicitly.

5.6   The Re-identification Risk of an Entire Microdata Set

In this section we present a rather crude model to evaluate the re-identification risk of an entire microdata set, i.e. the probability that any person in the microdata set can be re-identified. This model is described in [79]; see also [49] for an elaboration of this model.

Let the size of the population be denoted by N. We assume that a sample of size n has been drawn from this population. Let the sample fraction be denoted by $f(= n/N)$. For each sample element a number of variables is measured. The values obtained by these measurements are collected in records, one for each sample element. Together, the records constitute a data set S that will be made available to an intruder I. It is assumed that the variables in the key are categorical. The key can be considered as one variable with the possible number of values equal to the product of the possible numbers of values of the variables it consists of. Let k be the number of possible values of the key. Let f_u be the fraction of unique elements in the population. With respect to intruder R and key K a circle of acquaintances is assumed to be associated, consisting of persons of whose scores R knows with respect to K, or consisting of another microdata set available to R that R uses to identify individuals (*i.e.* an identification file). The fraction of acquaintances of R in the population is denoted by f_a.

It is assumed that if conditions C1, C2 and C3 of the conditions for re-identification given in Section 2.3 hold, then conditions C4, C5 and C6 hold as well. Condition C4 is a rather exacting one, but it can be introduced as an assumption for the sake of convenience in formulating a disclosure risk model. Note that it then yields a worst-case situation, in the sense that fallible perception and memory or other sources of ignorance, confusion and uncertainty for a potential discloser are excluded (in case the intruder does not use an identification file). C4 taken as an assumption together with C5 and C6 the implication is that the occurrence of any unique acquaintance E of R in data set S is equivalent to re-identification by I. And one may assume in turn that re-identification can automatically lead to disclosure.

Another assumption is adopted, in order to avoid additional complications: none of the data involved contain errors, neither the data in the microdata set released nor the data used by the intruder to identify individuals. Clearly, this is a rather unrealistic assumption, which however is made for convenience. The model could be refined later by taking these complications into consideration. We briefly come back to this issue below.

The disclosure risk for a certain microdata set S with respect to a certain intruder R and a certain key K, is defined to be the probability that the intruder makes at least one disclosure of a record in S on the basis of K. Actual disclosure requires full completion of the set of events C1–C6 from Section 2.3. The probability of disclosure for given R and E can be denoted by $\Pr[C1, \ldots, C6]$. Instead of this probability model involving 6 events we consider a simplified one involving only the first three events in this expression, that is we consider:

$$D_R = \Pr[C1, C2, C3], \tag{5.17}$$

instead of $\Pr[C1, \ldots, C6] = D_R \times \Pr[C4, C5, C6 | C1, C2, C3]$. The last factor of this expression is the (conditional) probability that, provided that conditions C1–C3 are fulfilled, intruder R actually achieves full disclosure by completing C4–C6. This conditional probability, however, depends on rather subjective circumstances and is therefore much more difficult to model and estimate. If C4–C6 are assumed to be automatically fulfilled, then this conditional probability is trivially equal to one, so that the risk R is equal to the probability 5.17. On the whole, however, the risk R can be seen as an upper bound for the actual disclosure risk for intruder I.

In addition to C1–C6 it is assumed that the properties being an element of the sample (*i.e.* an element of D), being an acquaintance of R (*i.e.* being an element of A) and being unique in the population (*i.e* being and element of U), all considered as random events, are independent of each other. As a consequence, the expected number of elements in D, that are known by R and that are unique in the population is $N f f_a f_u$. By a simple reasoning it can be shown that in many cases we have that

$$D_R \approx 1 - e^{-Nf f_a f_u}. \tag{5.18}$$

In equation 5.18 we have the unknown parameters f_a and f_u. The parameter f_a can be "guestimated" (*i.e.* obtained by inspired guesswork) by assuming a suitable scenario for an intruder. A number of such scenarios has been described in [52] and [53]. Contrary to the inspired guesswork to find a plausible value for f_a, the estimation of f_u is a genuine statistical problem. For the estimation of f_u, various models have been proposed in the literature, such as the Poisson-gamma model, the negative binomial model, the Poisson-lognormal model, models based on equivalence classes, and others (*cf.* [3], [13], [40], [41], [64] and [67]). A problem closely related to the estimation of f_u is that of determining the number of scores that appear in the population for a given key. In the biometry literature this problem is studied under the heading of species abundance models (*cf.* [33]). Each species corresponds to a particular combination of scores on a given key.

In terms of the Paaß and Wauschkuhn (*cf.* [52], p. 162) set-up f_a and f_u together reflect the *Informationsgehalt der Überschneidungsmerkmale*, *i.e.* the information content of the matching values. The various scenarios they consider differ in terms of f_a and f_u. In particular, f_u is influenced by the number of variables and the information content of these variables, *i.e.* their categorization, an attacker has at his disposal to re-identify a record. The parameter f_a is determined by the number of records that are contained in the identification file.

Comparing the approach sketched above to Paaß and Wauschkuhn's (*cf.* [52]) and Skinner's *et al.* (*cf.* [64]) approach it is apparent that they allow for some of the values in the microdata set S to be distorted, due to various error sources (response errors, coding and typing errors, *etc.*). This introduces a matching error when an intruder wants to re-identify an individual from D. This matching error was excluded from the above model, since condition C6 was not incorporated into the approach in this section. As was remarked before the model presented above can be extended to include this aspect as well. A major problem in practice, however, seems to be the estimation of this matching error, in particular the amount of knowledge about this error that can be attributed to an intruder.

From equation 5.18 it is clear that the statistical office that disseminates the data is able to influence the risk of re-identification by reducing f or f_u, assuming that it has no possibility to influence f_a (in [49], another point of view is presented, however). Reduction of f can be achieved through subsampling: instead of releasing the original file only a subsample thereof is released. This method is a bit wasteful as to the information collected.

A reduction of f_u can be achieved through global recoding[1] or local suppression applied to the key variables. The information loss achieved in this way is generally much less than the loss due to subsampling.

So far a theory is sketched to assess disclosure risks for a microdata set. Clearly such a theory alone is insufficient to cope with practical disclosure avoidance problems. What is needed in addition is a method, based on the theory developed, that can be used to decide whether a given microdata set is safe, and if not what actions should be undertaken to produce a safe microdata set from the given one. We proceed to sketch such a method, based on the model discussed in the present section.

Suppose that the data set S is not "safe" and that we want to take a subsample or collapse strata defined by the key. What is the effect on the disclosure risk function of either operation? To answer this question we consider the risk function as defined by the right-hand side of 5.18. Let R be the risk for D, R_1 the risk for a data set D_1 obtained from S by subsampling and R_2 the risk for a data set D_2 obtained from S by collapsing some strata of the key. We have

$$D_R = 1 - e^{-Nf f_a f_u} = 1 - e^{-\alpha n} = 1 - e^{-\beta U}, \qquad (5.19)$$

where $\alpha = f_a f_u$ and $\beta = f f_a$ and n is the size of D and U the number of population uniques with respect to a particular key of D. We assume that both α and β are small. Furthermore we have

$$R_1 = 1 - e^{-Nf' f_a f_u} = 1 - e^{-\alpha n'}, \qquad (5.20)$$

where n' is the size of D_1, and

$$R_2 = 1 - e^{-Nf f_a f_{u'}} = 1 - e^{-\beta U'}, \qquad (5.21)$$

where U' is the number of population uniques with respect to the key of D_2. Then

$$\frac{R}{R_1} = \frac{1 - e^{-\alpha n}}{1 - e^{-\alpha n'}} \approx \frac{n}{n'}, \qquad (5.22)$$

The approximation is justified because it was assumed that α is small. Similarly we find

$$\frac{R}{R_2} \approx \frac{U}{U'} \qquad (5.23)$$

Note that the reduction in risk through subsampling can be immediately calculated when the size of D_1 is known. In case of global recoding of some key variables this is more difficult, because U and U' have to be estimated.

---

[1] Strictly speaking *local recoding*, *i.e.* recoding some of the key variables for some of the records, would suffice, but this would produce a rather odd sort of file.

6
Tabular Data

Table talk, to be perfect, should be sincere without bigotry,
differing without discord, sometimes grave, always agreeable,
touching on deep points, dwelling most on seasonable ones,
and letting everybody speak and be heard.
—Leigh Hunt

6.1 Introduction

Somewhat abstractly speaking a table consists of a set of cells. Each cell is characterized by a set of coordinates, consisting of combinations of scores on certain categorical variables. Often these variables are identifying variables. To each cell there corresponds a cell total, consisting of the (possibly weighted) sum of individual contributions. Because the cell total is the result of an addition, the variable associated with the cell total has to be a numerical variable. Often the tabulation variable is a sensitive variable.

Examples of tables are the total profits of enterprises per economic activity and size class, or the total number of persons in a region per sex and age. In general, tables do not contain information about individual entities (*e.g.* respondents), but information about a collective, whose members have certain characteristics in common, namely those that identify the cell. However, situations may occur in which it is possible to deduce information corresponding to an individual entity from the aggregates collected in a table or set of tables. This can either be this individual's contribution to the cell, or a good approximation thereof. The task of SDC for tables is to avoid this.

In this chapter we discuss SDC of tabular data. The first step in this process will be the determination of the cell values that should not be published, because otherwise information from individual respondents may be disclosed. Methods to trace these so-called sensitive cells are discussed in Section 6.2 and Section 6.3. Section 6.2 deals with tables where the cells contain magnitude data, *i.e.* aggregates of a continuous variable, while Section 6.3 is devoted to frequency count data (*i.e.* where the cells are integer valued). A prerequisite for a cell to be sensitive is that it is the value of a sensitive variable. The reader should not confuse *sensitive cells* with *sensitive variables*: a sensitive cell requires a sensitive variable, but a sensitive variable does not necessarily yield sensitive cells.

After the sensitive cells have been determined in a table, measures have to be taken to conceal the sensitive information in these cells. A first measure is to redesign the table, by collapsing rows, columns, *etc.* in the table. This results in less detailed tables, with a reduced disclosure risk. With this SDC measure the information in the table is modified *globally*. Furthermore, measures can be taken which *locally* modify the table. For instance one might suppress the cell totals in the sensitive cells. These cell values are then set to *missing*. If a table contains marginal totals as well, these primary suppressions may not be sufficient because they can be easily re-calculated. So extra cells in the table have to be suppressed, the so-called secondary suppressions. Finding these is generally an arduous task, in particular if one has to consider the precision with which the suppressed cell values can be re-calculated. This may be the case if extra constraints on the possible values of the cells are taken into account, such as their nonnegativity. Another technique is to round the information in each cell such that only an approximation of the actual cell values is published. Section 6.4 deals with these techniques in more detail.

In Section 6.5 a procedure for the SDC of tabular data is described in which the results from the previous sections are combined. This section also contains a number of recommendations and remarks regarding SDC of tabular data. Furthermore, the disclosure control of tabular data based on samples is discussed. This is a setting one sometimes encounters in practice. The theory discussed in the previous sections, which assumed that the tables contained information on the entire population, is then still applicable except that one has to reconsider the concept of a sensitive cell.

So far only single tables (and their marginals) have been considered. In Section 6.6, the closing section of this chapter, we discuss a more general setting that is also encountered in practice: linked tables. Such tables are produced by different aggregate actions from the same base file, and they may possibly have variables in common. Linked tables not only occur regularly in practice, it also seems that they are the natural setting for discussing SDC for tables. The theory of producing safe (two-)dimensional tables is just a stepping stone to the more general theory of the production of safe linked tables. For practical purposes it is useful to make a distinction between fixed sets of linked tables and non-fixed sets. Sets of the latter type occur for instance when tables are produced and released in due course, without any preset plan of which tables are to be released.

6.2 Tables with Magnitude Data

The first step in the disclosure control procedure of tabular data consists of the determination of the sensitive cells, *i.e.* the cells that tend to reveal too much information about an individual respondent. The next step will be to avoid that after publication of the table it is still possible to disclose the exact value of the sensitive cell or, more generally, to make an estimate of this value which is unacceptably close. Throughout this section we will assume that the tables under consideration are based on an observation of the entire population.

Several criteria exist to determine whether a cell is sensitive and therefore should not be published, at least not in its present form. A commonly used measure is the *(n,k)-dominance rule* which states:

A cell is regarded sensitive if the sum of the largest n contributions account for more than k% of the total cell value.

Usually, a small value is taken for the parameter n, at most up to 5, while k is taken large, for instance 80. The main idea behind this rule is the following. One tries to avoid that $n-1$ respondents can make a good estimate of the contribution of the n-th respondent by pooling their own values and by comparing this sum with the total cell value. The larger n, the less likely that $n-1$ respondents will cooperate. Therefore, the parameter n should be chosen larger than the maximum size of (imagined) coalitions of respondents. Special cases are $n = 1$ and $n = 2$. In these cases the cell total is dominated by the contributions of one or two respondents, respectively. If a cell value is dominated by the contribution from one respondent, then this contribution can be estimated fairly accurately by anybody who reads the cell value. In particular, if there is only one respondent then his contribution will be disclosed exactly. Notice that dominance by a specific respondent, *e.g.* a firm, in a given cell of the table often is public knowledge. If the value of a cell is dominated by the contribution from two respondents (for instance, if these respondents make up 90% of the total cell value), then each of these respondents is able to estimate the value of the contribution of the other one rather accurately. In particular, if there are just two respondents, then each of these respondents can disclose the contribution of the other exactly by subtracting his own contribution from the total cell value. In view of these cases a dominance rule should prevent cells to be published which are dominated by one or two respondents. As a consequence, the minimum number of respondents corresponding to each cell should at least be 3. Finally, notice that in order to apply the dominance rule, one needs to know the cell total as well as the top k individual contributions to the cell total, where n is the first parameter of the dominance rule, specifying the critical coalition size (plus 1).

TABLE 6.1. Investments of enterprises × 1 million guilders

	Region A	Region B	Region C	Total
Activity I	20	50	10	80
Activity II	8	19	22	49
Activity III	17	32	12	61
Total	45	101	44	190

Table 6.1 gives an example to demonstrate the working of this rule.

We choose the following parameter values: $n = 3$ and $k = 75$. Suppose the individual contributions from the respondents corresponding to activity I and region B are, in descending order:

    enterprise 1   :   16,
    enterprise 2   :   10,
    enterprise 3   :   6,
    others (total)   :   18.

The sum of the largest three contributions is equal to 32, which is 64% of the total cell value. Therefore, this cell will not be regarded sensitive according to the dominance rule used. A coalition of two enterprises will not be able to make an accurate estimate of the contribution of a third enterprise.

Next, consider the contributions of the respondents corresponding to activity II and region C. Suppose these contributions are given by

    enterprise 1   :   8,
    enterprise 2   :   6,
    enterprise 3   :   5,
    others (total)   :   3.

In this case the largest three contributions add up to 19, which is 86% of the cell total of 22. If, for instance, enterprises 2 and 3 would cooperate, they could make an accurate estimate of the contribution from enterprise 1, namely between 6 and 11, provided they know that enterprise 1 has the largest contribution, which within a group of firms often applies. Anybody not included in this cell could make an estimate between 0 and 22 only, assuming he does not know the number of enterprises contributing to this cell. This cell total should not be published in its original form. A disclosure control measure is therefore necessary.

Apart from the dominance rule other rules for determining the sensitive cells have been suggested. An example of such a rule is the (p,q)-prior-posterior rule. This rule uses two parameters, p and q with $p < q$. It is assumed that every respondent can estimate the contribution of each other respondent to within q percent of its respective value. This "prior"-knowledge may be based on the general knowledge a respondent has about the population in question. After a table has been published, the information of the respondents changes ("posterior"-knowledge) and they may be able to make a better estimate of the contribution of another respondent. A cell is considered sensitive if on the basis of its value it is possible to estimate the contribution of an individual respondent to that cell to within p percent of the value of this contribution. Again a disclosure control measure is necessary to protect the sensitive cells. We refer to [10] and [38] for more information on this rule.

Other measures for the sensitivity of a cell value, besides the (n,k)-dominance rule or the (p,q)-prior-posterior rule, are possible. Such rules could for instance be based on Lorenz-curves or Gini-indices. A sensitivity measure at least should satisfy the condition that the sum of two non-sensitive totals is again a non-sensitive total. This property is called sub-additivity. In [10] linear sensitivity measures are discussed in general, while in [75] linear subadditive sensitivity measures are studied.

6.3   Sensitive Cells in Frequency Count Data

The discussion in Section 6.2 refers to tables with quantitative data, like the turnover or the profits of enterprises. One might be tempted to think that the disclosure control of tables with frequency count data is in principle the same as for tables with magnitude data. Mechanically applying the dominance rule yields for frequency count data that small cell values, for instance cells with less than three respondents, should not be published. However, in many cases a dominance rule is not the right measure to determine whether a cell value should be considered sensitive. When a table does not contain any sensitive variables, then the application of a dominance rule is not appropriate. On the other hand, when a table does contain sensitive information about the respondents, then the application of a dominance rule may not be adequate. In this section we will discuss these two points in detail. We start by examining Table 6.2 which is in fact part of a larger table on environmental offenses by firms.

It becomes evident from the information in Table 6.2 that the only firm in region B with activity III has committed an environmental offense. Anyone reading this table is able to arrive at this conclusion, about a uniquely specified firm. A weaker, but possibly still inadmissible, form of disclosure is illustrated in Table 6.3.

Here the only firm in region B with activity III which did not commit an environmental offense arrives, on the basis of Table 6.3, at the conclusion that all his "companions" did. An outsider can conclude about a given one of these 5 firms only that there is a probability of 80% that it has committed such an offense. Each of the offenders can conclude the same

TABLE 6.2. Environmental offense by Activity and Region

Environmental offense	Yes	No	Total
...	...	...	...
Activity III, Region B	1	0	1
...	...	...	...

TABLE 6.3. Environmental offense by Activity and Region

Environmental offense	Yes	No	Total
...	...	...	...
Activity III, Region B	4	1	5
...	...	...	...

about their "companions" with a probability of 75%.

Table 6.4 combines the features of both foregoing tables. In this case there are many offenders and no firm is free from guilt. Both insiders and outsiders know with certainty on the basis of this table that each of the 12 firms has committed an environmental offense. There is no question of disclosure on the basis of unicity with respect to identifying variables. Disclosure takes place because *all* firms in the group committed an environmental offense. Such a situation is termed *group disclosure*.

TABLE 6.4. Environmental offense by Activity and Region

Environmental offense	Yes	No	Total
...	...	...	...
Activity III, Region B	12	0	12
...	...	...	...

Finally, we present a table, Table 6.5, from which one can read not only that each member of the group concerned has committed an environmental offense, but also that a slight majority of them drained polluted sewage.

From these examples it is clear that in the case of frequency count data the dominance rule from Section 6.2 is not the right means to deal with

TABLE 6.5. Environmental offense by Activity and Region

	Draining polluted sewage	Other environmental offense	No environmental offense	Total
...	...	...	...	
Activity III, Region B	5	4	0	9
...	...	...	...	

this kind of problems. Rather, one would have to direct the attention to the distribution on the various categories of the variable in question. More specifically, if the (majority of the) population scores on really sensitive categories only, like in the Tables 6.2 through 6.5, then publication of that information is not allowed. A sufficiently large part of the population should score on the non-sensitive categories.

In principle, the situations illustrated by Tables 6.2 through 6.5 may occur for magnitude data as well. For instance Table 6.6 shows the tons of polluted sewage that have been disposed of in a legal and in an illegal way.

TABLE 6.6. Disposed polluted sewage (in tons)

	Legally disposed	Illegally disposed	Total
...	...	...	...
Activity III, Region B	0	11.2	11.2
...	...	...	...

From Table 6.6 we can conclude that all the companies in region B with activity III that disposed polluted sewage did this in an illegal way. When, moreover, activity III is such that polluted sewage must be disposed of one way or the other, we can conclude that all the companies in region B with activity III disposed polluted sewage in an illegal way.

It is clear, however, that the situations illustrated by Table 6.2 through 6.5 are much more likely to occur for frequency data than for magnitude data.

6.4   Disclosure Control Measures

The first part in the disclosure control procedure of tabular data has been formulated in the previous sections as the determination of the sensitive cells, *i.e.* the cells that cannot be published unaltered. In this section we discuss three methods to conceal the sensitive information in the tables, indicating at the same time the differences between the methods.

6.4.1   Table Redesign

It is possible that a lot of cells in a row or column appear to be sensitive. In that case it is recommended that first the table is redesigned, *i.e.* that its classification scheme is modified. In this way the detail of the statistical information is reduced. The aim is that the number of sensitive cells decreases due to a reduction of detail in the table.

We will illustrate this technique by means of a simple example. Consider Table 6.7 (the same as Table 6.1) shows the investments of enterprises classified by region and by activity of the enterprises.

TABLE 6.7. Investments of enterprises × 1 million guilders

	Region A	Region B	Region C	Total
Activity I	20	50	10	80
Activity II	8	19	22	49
Activity III	17	32	12	61
Total	45	101	44	190

Suppose that most of the cells related to the activities II and III are considered to be sensitive. As a disclosure control measure the rows corresponding to these activities are combined. The result is presented in Table 6.8.

TABLE 6.8. Investments of enterprises × 1 million guilders (after table redesign)

	Region A	Region B	Region C	Total
Activity I	20	50	10	80
Activity II + III	25	51	34	110
Total	45	101	44	190

In the new table we again have to check whether the cells in the table satisfy the dominance rule. Additional disclosure control measures have to be taken for cells that are still sensitive after the aggregation of the two rows. For instance, a further reduction of detail is possible, in particular if the number of sensitive cells is still large. When there are a few sensitive cells left only, a local measure, like rounding or cell suppression, is recommended. These techniques are discussed in the Sections 6.4.2 and 6.4.4.

6.4.2 Cell Suppression

The first local method we discuss concerns the identification and suppression of the sensitive cells. The value of the sensitive cell is deleted and replaced by for instance the symbol ×. Those cells are called primary suppressions. Let us consider again the information presented in Table 6.7. Let us assume that the cell value corresponding to activity II and region C cannot be published in its present form according to the dominance rule used. This cell value is suppressed. Table 6.9 gives the results.

TABLE 6.9. Investments of enterprises × 1 million guilders (after primary suppression)

	Region A	Region B	Region C	Total
Activity I	20	50	10	80
Activity II	8	19	×	49
Activity III	17	32	12	61
Total	45	101	44	190

In general it will not be sufficient to suppress the sensitive cells only, as will be clear from Table 6.9. The suppressed cell in Table 6.9 can easily be computed by means of the marginal totals. A first suggestion to make the table safe for release could be to delete those marginal totals. However, the loss of information due to deleting the marginal totals may not be acceptable for the users of the table. The condition for publishing these marginal totals, of course, is that the marginal totals themselves are not sensitive. Another option is to suppress additional, non-sensitive internal cell values in the table. These suppressions are called secondary suppressions. A solution, but certainly not the only one for Table 6.9, is to suppress the values corresponding to activity II and activity III in region A, and to activity III, region C (cf. Table 6.10).

From Table 6.10 it is no longer possible to re-compute the value of the sensitive cell (activity II, region C) exactly.

In a larger table with a number of primary suppressions the choice of the secondary suppressions soon becomes a complex task when the aim is to minimize the information loss (suitably quantified). The following aspects play an important role:

TABLE 6.10. Investments of enterprises × 1 million guilders (after primary and secondary suppressions)

	Region A	Region B	Region C	Total
Activity I	20	50	10	80
Activity II	×	19	×	49
Activity III	×	32	×	61
Total	45	101	44	190

a. The sensitive cells should be adequately protected by the choice of the secondary suppressions; the ranges in which the values of the suppressed cells lie, should not be too narrow. It should be borne in mind that calculating ranges is possible when the values of the cells are restricted in some way, *e.g.* by a requirement that they be nonnegative.

b. The loss of information due to the secondary suppressions should be minimized.

c. No zero-valued cells or empty cells should be suppressed.

In the remainder of this section we examine these aspects in more detail.

Ad a.:
From the data in Table 6.10 we conclude that the investments by enterprises in region C with activity II add up to a value between 5 and 30 million guilders, where we assume that all cell values are nonnegative. No smaller range can be found from Table 6.10. Because this range is quite large, the protection provided by the secondary suppressions seems adequate enough. In other situations however, it is possible to obtain a small range within which the sensitive cell value has to lie. For example, in Table 6.11 the cells (1,1) and (2,1) are the primary suppressions, while the cells (1,3) and (2,3) are the secondary suppressions. The values of the suppressed cells are indicated with x_i, i=1,...,4.

From Table 6.11 the following relations can be deduced: $x_1 + x_2 = 103$ and $x_2 + x_4 = 4$. If it is also known that every cell value is nonnegative, then we can conclude that x_1 lies within the range [99,103]. Such a small range will in many cases be considered unacceptably narrow. When an attacker would use the average of 99 and 103, *i.e.* 101, as an estimate for x_1, then the difference between this estimate and the real value is less than 2%. In this case an attacker is able to make a good estimate of the sensitive cell value and, as a consequence, is probably able to disclose the contribution of an individual contributor. In order to make the table safe for release

TABLE 6.11. Table with small ranges for the suppressed cells

	C1	C2	C3	Total
R1	x_1	1	x_2	104
R2	x_3	2	x_4	103
R3	70	3	2	75
Total	270	6	6	282

other secondary suppressions have to be chosen. For instance, some of the marginal totals could be suppressed.

Such a range in which the values of a suppressed cell must lie will be called the *feasibility interval* for this cell. The feasibility intervals for the sensitive cells should be sufficiently wide. In the above example the feasibility interval of x_1 is [99,103]. The SDC rules should prescribe an interval, the *protection interval*, that should be contained by the feasibility interval corresponding to the sensitive cell. Instead of prescribing a particular protection interval the SDC rules frequently prescribe only the length of this interval. In the above example the SDC rules could prescribe that the length of the protection interval is 10. Because the length of the feasibility interval is only 4, the table is insufficiently protected.

In general, it will not be possible to find lower and upper bounds for the suppressed cells if the cells can assume both positive and negative values. Section 6.4.3 illustrates how the choice of the secondary suppressions influences the ranges in which the suppressed cell values lie.

General guidelines on how to choose the secondary suppressions such that no accurate estimates of the suppressed cell values are possible, seem hard to give. A reason for this is that the ranges within which the suppressed values lie, depend on the actual cell values. It is possible to give a necessary condition. The number of suppressed cells in each row or column of the table should be equal to zero or should at least be equal to two. When there is only one suppressed cell in a row or column, then its value can directly be computed using the marginal totals. Unfortunately this condition is not sufficient, as is shown in the following example. Consider Table 6.12, in which the suppressed cells are indicated by x_i, i=1,...,9.

From Table 6.12 the following relations can be deduced:

$$x_1 + x_3 + x_4 + x_5 = 12,$$
$$x_1 + x_2 + x_3 + x_4 + x_5 = 15.$$

From these relations it immediately follows that $x_2 = 3$. So, although in each row and column at least two cells are suppressed, it remains possible

TABLE 6.12. An insufficiently protected table

	C1	C2	C3	C4	Total
R1	x_1	x_2	x_3	2	13
R2	x_4	2	x_5	7	13
R3	3	x_6	8	x_7	14
R4	4	x_8	3	x_9	18
Total	11	10	19	18	58

to disclose the exact value of x_2.

Ad b.:

Different choices of the secondary suppressions will lead to different losses of information. In order to quantify the loss of information we assign a weight w_{ij} to each cell (i,j). Several choices for the weights are possible, dependent on the criterion which should be minimized, of which we mention:

- Minimize the total number of suppressed cells. In this case all cells are assigned an equal weight, in other words: $w_{ij} = c$, with $c > 0$.

- Minimize the total value of the suppressed cells. In this case the weight w_{ij} of cell (i,j) is chosen equal to the value d_{ij} of cell (i,j).

- Minimize the number of respondents corresponding to the suppressed cells. If N_{ij} denotes the number of respondents contributing to the value of cell (i,j), then the weight w_{ij} is chosen equal to N_{ij}. The cell values given in Table 6.13 could for instance be used in order to find secondary suppressions for Table 6.9.

TABLE 6.13. Number of enterprises by region and activity

	Region A	Region B	Region C	Total
Activity I	8	7	6	21
Activity II	4	5	4	13
Activity III	6	5	4	15
Total	18	17	14	49

Primary suppressions are assigned a weight equal to a large negative number; they have to be suppressed anyway. Besides the options mentioned above, other criteria are possible as, for instance, combinations of the given ones. It may be possible that for some reason a certain cell value should not be suppressed. In that case a large positive weight is assigned to such a cell.

We now aim to choose the secondary suppressions in such a way, that the loss of information according to the criterion used is minimal under the condition that the chosen secondary suppressions adequately protect the sensitive cells. A solution which is optimal according to a certain criterion may not be optimal according to another criterion. In Chapter 4 of [38] these criteria are discussed in more detail.

Next, we will give an example to show that different choices for the weights w_{ij} may lead to different secondary suppressions. Table 6.14 gives the total number of employees in enterprises by economic activity and size (in classes) of those enterprises. The cells with economic activity 6 and size classes 5 and 7 appeared to be sensitive and have been suppressed.

TABLE 6.14. Number of employees by economic activity and size

	SC 4	SC 5	SC 6	SC 7	SC 8	Total
EA 2,3	80	253	54	-	-	387
EA 4	641	3,694	2,062	746	-	7,143
EA 5	592	88	329	1,449	1,440	3,898
EA 6	57	×	946	×	2,027	4,281
EA 7	78	-	890	1,719	1,743	4,430
Total	1,448	4,353	4,281	4,847	5,210	20,139

If we choose to minimize the number of secondary suppressions, meaning that we take the same weight for all cells, we obtain the secondary suppressions given in Table 6.15.

TABLE 6.15. Secondary suppressions when $w_{ij} = c$

	SC 4	SC 5	SC 6	SC 7	SC 8	Total
EA 2,3	80	253	54	-	-	387
EA 4	641	×	2,062	×	-	7,143
EA 5	592	88	329	1,449	1,440	3,898
EA 6	57	×	946	×	2,027	4,281
EA 7	78	-	890	1,719	1,743	4,430
Total	1,448	4,353	4,281	4,847	5,210	20,139

A consequence of this choice is that a cell with a large number of employees (activity 4 and size class 5) is suppressed. This feature might be

avoided, if we suppress the values of the cells with activity 5 and size class
5 and 7 instead. In Table 6.16 the secondary suppressions are given when
we minimize the total value of the secondary suppressions, *i.e.* $w_{ij} = d_{ij}$.

TABLE 6.16. Secondary suppressions when $w_{ij} = d_{ij}$

	SC 4	SC 5	SC 6	SC 7	SC 8	Total
EA 2,3	×	×	54	-	-	387
EA 4	×	3,694	2,062	×	-	7,143
EA 5	592	88	329	1,449	1,440	3,898
EA 6	×	×	946	×	2,027	4,281
EA 7	78	-	890	1,719	1,743	4,430
Total	1,448	4,353	4,281	4,847	5,210	20,139

In this case especially cells with a low number of employees are suppressed. A consequence of this choice is that Table 6.16 contains five secondary suppressions, whereas Table 6.15 contains only two.

As a last example we give the secondary suppressions when we want to
minimize the total number of contributors corresponding to the cells that
have been suppressed. In order to do so, we need as extra information the
number of enterprises N_{ij} which contribute to cell (i,j). These numbers
are given in Table 6.17, while the resulting secondary suppressions can be
found in Table 6.18.

TABLE 6.17. Number of enterprises per cell (N_{ij}).

	SC 4	SC 5	SC 6	SC 7	SC 8
EA 2,3	10	8	3	-	-
EA 4	60	100	40	11	-
EA 5	55	5	14	22	9
EA 6	7	2	22	2	16
EA 7	9	-	18	30	12

Again, the number of secondary suppressions is equal to five. Compared
to Table 6.16 however different cells have been suppressed. We mention
here that in all cases we have taken care not to suppress empty cells (see
Ad c., below, for an explanation). We did not take into account whether
the secondary suppressions are adequate. This depends on the purpose of
the table, which table should finally be published. At least it is necessary
to avoid that in different publications tables with different secondary suppressions appear. For example, when both Table 6.16 and Table 6.18 are
published, it is possible to obtain additional information about the values

TABLE 6.18. Secondary suppressions when $w_{ij} = N_{ij}$

	SC 4	SC 5	SC 6	SC 7	SC 8	Total
EA 2,3	×	×	54	–	–	387
EA 4	641	3,694	2,062	746	–	7,143
EA 5	592	×	329	×	1,440	3,898
EA 6	×	×	946	×	2,027	4,281
EA 7	78	–	890	1,719	1,743	4,430
Total	1,448	4,353	4,281	4,847	5,210	20,139

of the sensitive cells by combining the information from the two tables.

Ad c.:
Within the population of enterprises it may be a well-known fact that
there are no enterprises with a given combination of characteristics. The
cell value corresponding to this combination is therefore equal to zero. An
example can be found in Table 6.14, in which the cell corresponding to
economic activity 4 and size class 8 is empty. If such a cell would be chosen
as a secondary suppression, then it may be possible, in combination with
the information that this cell is empty, to disclose the value of another
suppressed cell. To suppress an empty cell does not conceal anything in
this case.

6.4.3 Feasibility Intervals

In the previous section we indicated that the secondary suppressions should
be chosen in such a way that not only exact re-calculation of suppressed
values is excluded, but also that no unacceptably accurate estimates are
possible. In this section we briefly illustrate how the choice of the secondary
suppressions influences the *feasibility intervals*, *i.e.* the ranges in which the
values of the suppressed cells must lie.

Consider Table 6.10 again. For this choice of the secondary suppressions
the feasibility intervals for the suppressed cells are indicated in Table 6.19.

TABLE 6.19. Investments of enterprises × 1 million guilders (with feasibility intervals for suppressed cells)

	Region A	Region B	Region C	Total
Activity I	20	50	10	80
Activity II	0–25	19	5–30	49
Activity III	0–25	32	4–29	61
Total	45	101	44	190

If we would have chosen other secondary suppressions, then also the feasibility intervals for the suppressed cells change. Consider for instance Table 6.20 in which other cells are suppressed in order to protect the sensitive cell corresponding to activity II and region C.

TABLE 6.20. Investments of enterprises × 1 million guilders (after primary and secondary suppressions)

	Region A	Region B	Region C	Total
Activity I	20	×	×	80
Activity II	8	×	×	49
Activity III	17	32	12	61
Total	45	101	44	190

Again we can determine the feasibility intervals for the values of the suppressed cells. The result is presented in Table 6.21.

TABLE 6.21. Investments of enterprises × 1 million guilders (with feasibility intervals for suppressed cells)

	Region A	Region B	Region C	Total
Activity I	20	28–60	0–32	80
Activity II	8	9–41	0–32	49
Activity III	17	32	12	61
Total	45	101	44	190

From Table 6.21 we learn that the value of the cell corresponding to activity II and region C is in the interval [0,32]. From Table 6.19 we concluded that the feasibility interval was equal to [5,30]. We conclude that the suppression pattern in Table 6.20 offers a better protection to the sensitive cell than the suppression pattern in Table 6.10. A drawback is that in Table 6.20 cells with larger values are suppressed than in Table 6.10. As a result, the information content of Table 6.20 may be less than that of Table 6.10.

6.4.4 Rounding

A third disclosure control measure to safeguard tabular data is rounding the cell values to integer multiples of a fixed base. Rounding can be seen as a special way of perturbing the data. In both cases the data are changed in order to protect the underlying information.

Several methods exist for rounding tabular data. The easiest one is conventional rounding, where each value is rounded to the nearest multiple of the rounding base. Conventional rounding has two serious limitations. Firstly, usually the additive structure of the table is not preserved. A simple example will make this clear. Consider a one-dimensional table with 2 internal cells which both have the value 2. Hence, the marginal total equals 4. After rounding with 5 as rounding base, both the internal cells have the value 0, while the marginal total is equal to 5. A second limitation, which is important from the point of view of SDC, is that it is sometimes possible to disclose the original cell values or to restrict the number of possible "original values" considerably. From the previous example this becomes immediately clear. The only way in which two numbers which both are rounded to zero, can add up to a number which is equal to 5 after rounding is when one of the numbers equals 2, and the other one equals 1 or 2.

In order to overcome the limitations of conventional rounding, other techniques have been developed. In [35] and [11] so-called controlled (random) rounding procedures are suggested for one- and two-dimensional tables, respectively. These methods are explained in detail in Section 7.4. With both procedures it is possible to preserve the additive structure of the table: the sum of the rounded cell values is equal to the rounded sum of the cell values. The procedures consist of rounding the unrounded cell values, including the totals, to one of the two adjacent multiples of the rounding base, instead of rounding the value to the nearest multiple. Values that are already an integer multiple of the rounding base are not modified. To which one of the two adjacent multiples of the rounding base an unrounded value is rounded is determined by a simple random process. Due to this random process the procedures are unbiased: the expected outcome of the rounded value is equal to the original value.

A practical aspect of rounding is the choice of the rounding base. Rounding should provide the sensitive cells with a sufficiently wide feasibility interval. Unacceptably close estimates of those cells should not be possible. Therefore, the rounding base should be chosen larger than a certain minimum value which depends on (the contributions to) the values of the sensitive cells. On the other hand the rounding base should not be chosen too large, since in that case too much information contained in the non-sensitive cells is lost.

Controlled rounding of higher-dimensional tables while at the same time preserving the additivity of the table seems to be quite difficult. One of the reasons is that there are three-dimensional tables in which it is impossible to round each unrounded value to one of the two adjacent multiples of the rounding base while the rows and columns still add up to their rounded marginal totals. Causey, Cox and Ernst (*cf.* [6]) display a simple example of such a three-dimensional table. In a slightly adapted form this example is given in Section 7.4.

Some heuristics to deal with three-dimensional tables have been developed at the U.S. Bureau of the Census (see [34]). Also in [44] heuristics are developed. In the same paper, as well as in [43], a mixed integer programming formulation for the three-dimensional controlled rounding problem is given. For four- and higher-dimensional tables satisfactory heuristics seem hard to find.

6.4.5   *Averages and Fractions*

In Section 6.4 emphasis is on aggregates like totals and frequencies. Averages and fractions constitute an equally important class of tabular data. There are some differences, of which the most conspicuous one is the absence of additivity within rows and columns. Because of the fact that totals (frequencies) and averages (fractions) are directly related to one another via the corresponding population numbers, which mostly are well-known on a large scale, SDC is the same in these cases. Hence we need not enter the aspects of disclosure control of tabular data of averages and fractions any further.

6.5   Recommendations

6.5.1   *A Procedure for SDC of Tabular Data*

In this section we describe a procedure for SDC of tabular data. The procedure combines methods from the previous sections. We refer to these sections for further details about the methods mentioned in this procedure. The first step invariably is the determination of the sensitive cells. We start by considering magnitude data.

Magnitude Data

- **Step 1**: *Determination of the sensitive cells.*
  Cell values which cannot be released usually are determined by a
  dominance rule. The following choices for the parameter values n
  and k in such rules are recommended: $n \geq 2, k \leq 80$.

- **Step 2**: *Table redesign.*
  Collapse some of the rows and/or columns which contain relatively
  many sensitive cells and construct the table with the new classification. Check whether the cells in the new table are sensitive or not.
  When still relatively many cells are sensitive, it is recommended to
  further aggregate the variables. Otherwise local SDC measures can
  be taken (*cf.* step 3a or step 3b).

Next, suppress or round the remaining sensitive cells and, if necessary,
adapt additional cells in the table to protect the sensitive cells.

- **Step 3a**: *Suppression.*
  Suppress the sensitive cells and, if necessary, additional cells. Check
  whether the feasibility intervals of the sensitive cells are wide enough.

  Rules to check whether the feasibility intervals of the sensitive cells
  are sufficiently wide can be constructed in several ways:

  1. The simplest option is to demand that the width of the feasibility
     interval of a sensitive cell is at least $\ell_{tot}\%$ of the real value of
     this cell. In this way the total cell value cannot be estimated
     accurately. As a consequence no contribution to the total cell
     value can be estimated accurately.

  2. The parameters of the dominance rule can used to determine
     the upper bound of the protection interval for a sensitive cell
     (*cf.* [9] and [39]). One can demand that at least one of the val-
     ues in the feasibility interval of the sensitive cell does not violate
     the (n,k)-dominance rule. That is, the upper bound of the fea-
     sibility interval should not violate the (n,k)-dominance rule. In
     this case, it is not allowed that n or fewer respondents account
     for $k\%$ or more of the upper bound of the feasibility interval. In
     other words, the upper bound U_F of the feasibility interval of a
     sensitive cell should at least be equal to $(100/k) \sum_{i=1}^{n} y_i$, where
     y_i is the i-th largest contribution to this cell. However, determin-
     ing a lower bound L_F for the feasibility interval of a sensitive
     cell is more difficult. A simple solution is to make the feasibility
     interval approximately symmetric around the real value.

     When one wants to prescribe the width of a feasibility interval
     of a sensitive cell rather than its upper and lower bound, then
     one can simply take

$$2 \times \left((100/k) \sum_{i=1}^{n} y_i - \sum_{i=1}^{N} y_i \right), \qquad (6.1)$$

where N is the total number of contributions to the cell, as the minimal width. That is, one can take the width of the (approximately) symmetric interval mentioned above as the minimal width of a feasibility interval.

3. The largest contribution to the total value of a sensitive cell can be used to determine the minimal width of a feasibility interval. One can demand that the width of the feasibility interval of a sensitive cell is at least $\ell_{max}\%$ of the largest contribution to this cell. The largest contribution, and consequently any other contribution, to a sensitive cell cannot be estimated more accurately than to within $\ell_{max}/2\%$ of the largest contribution.

To determine the minimal width of the feasibility interval of a sensitive cell only (the contributions to) the total cell value have to be considered.

- **Step 3b**: *Rounding.*
  Round the cell values to multiples of a previously chosen basis. By choosing the rounding basis sufficiently large, the sensitive cells acquire the appropriate protection.

Again rules to determine the value of the rounding base b can be constructed in several ways:

1. The rounding base b can be chosen to be at least $\ell_{tot}\%$ of the maximal total value of an internal sensitive cell in the table. As a result, the values of the sensitive cells cannot be estimated more accurately than to within $\ell_{tot}/2\%$ of the maximal total value of an internal sensitive cell in the table.

2. The parameters of the dominance rule can be used to determine the rounding base. For instance, the minimal value of the rounding base b can be chosen equal to

$$\max_{j} \left\{ (100/k) \sum_{i=1}^{n} y_{ij} - \sum_{i=1}^{N} y_{ij} \right\}, \qquad (6.2)$$

where y_{ij} denotes the i-th largest contribution to sensitive cell j. The maximum is taken over all sensitive cells j.

3. The largest contribution to the values of the sensitive cells can be used to determine the rounding base b. One can demand that the rounding base is at least $\ell_{max}\%$ of the largest contribution

to the sensitive cells. The largest contribution, and consequently any other contribution, to the sensitive cells cannot be estimated accurately.

To determine the minimal value of the rounding base of a table all (contributions to the) total cell values of the sensitive cells have to be considered.

A final, and important, remark is that the actual choice of the parameters n, k and ℓ should never be published or otherwise be made available (*cf.* Section 6.5.2).

Frequency Count Data

With frequency count data one has to consider the entire distribution over the categories of the variable simultaneously, and not only the various cells separately. Make sure that the counts are not concentrated in the really sensitive categories. Aggregation might not help in these cases, but rounding and suppression could be used, for which the demands are the same as in the steps 3a and 3b above.

6.5.2 Warning

Knowledge of the parameter values, or additional information in general, is another problem when reducing the risk of disclosure for a particular table. For example, suppose that the following dominance rule is used: a cell is regarded sensitive if at least 80% of its value is the combined result of the data of 2 companies. Suppose, furthermore, that, as up to now, all companies are in the survey, that there are three companies contributing to a certain cell total and that the largest company contributes 50% to the value of this cell. If the cell value is published and if the largest company happens to know the parameter values of the dominance rule (and that it is indeed the largest contributing company!), then it can deduce that the contribution of the second largest company to the cell value lies between 25% and 30% of the published total value of the cell. On the other hand, if the cell value is suppressed and if again the largest company knows the parameter values, then it can deduce that the suppressed value is not larger than 2.5 times its own contribution, because this contribution is at least 40% of the cell value. Both cases of estimation might be undesirable. Absolute secrecy about the parameter values of the dominance rule is the first step to avoid these problems.

6.5.3 Tabular Data Based on Samples

The foregoing sections bear upon the situation that the survey from which the data are derived is a complete enumeration or census. However, in most cases data are derived from a sample survey, in which only a part of the complete population is covered. Another complication is due to the occurrence of non-response, which also causes the exclusion of some members of the target population from the survey. These aspects imply that in estimating population parameters each observation has to be weighted with an appropriately chosen factor. These factors are based on the sampling scheme, the nature and the extent of the non-response and the procedure of estimation. Hence, disclosure of individual scores from totals or frequency counts is more complicated in such cases than it is in the case of a complete enumeration of the population.

The first factor counteracting disclosure is the fact that it is in general not known to the user of the tables which elements of the population have been included in the sample. Hence, SDC is favored by keeping the identity of the respondents secret. For surveys among enterprises this is not always feasible: it might be public knowledge that the large firms are included in the survey with absolute certainty, because otherwise the accuracy of the estimates of, for instance, totals of turnover would be too low. As far as the weights are concerned, the situation is different. In many cases the weighting scheme is of great value for an adequate use of the resulting aggregate values, and therefore should be released simultaneously with the tabular data.

In view of the considerations just given it is advisable to proceed in the case of samples in the same way as in the case of a complete enumeration of the population. Simple examples show that knowledge about which units contributed to a cell total, in combination with knowledge on the corresponding weights, can be of great help in disclosing individual contributions in the case of aggregates which are based on a small number of units. So, in publishing totals dominance rules should be applied at the level of the sample. In case one wants to suppress some cell values one generally has to make one or more secondary suppressions. Only if one can be reasonably certain that the users have no information about the weighting scheme and about the identity of the participants in the survey, one might relax the SDC measures. For instance, in tables with averages the relation between the cell values in the margin and those in the corresponding row or column is established via weighting. If these weights are not known to the user, primary suppressions might be sufficient for keeping sensitive cell values secret.

6.6   Linked Tables

It is possible that a table by itself does not contain any sensitive cell, but that by combining the information of that table with information from other tables it is still possible to disclose information about a respondent. We will illustrate such a situation by means of the following example.

A survey among self-employed shopkeepers in a certain town yields information about the location of business in the town (indicated by the center or the outskirts), the sex of the shopkeepers and the financial position (indicated by weak or strong). Three tables are published, the Tables 6.22, 6.23 and 6.24.

TABLE 6.22. Location of business and sex of self-employed shopkeepers

	Center	Outskirts	Total
Male	19	6	25
Female	3	6	9
Total	22	12	34

TABLE 6.23. Financial position and sex of self-employed shopkeepers

	Weak	Strong	Total
Male	13	12	25
Female	3	6	9
Total	16	18	34

TABLE 6.24. Location of business and financial position of self-employed shopkeepers

	Weak	Strong	Total
Center	7	15	22
Outskirts	9	3	12
Total	16	18	34

From these three tables it can be deduced that the financial position of all male shopkeepers in the outskirts of the town is weak. This disclosure is achieved by the following reasoning.

$$N_S(\text{male, outskirts}) = N_S(\text{male, outskirts, weak}) + N_S(\text{male, outskirts, strong}),$$

where N_S denotes the number of shopkeepers. For example, $N_S(\text{male, outskirts})$ denotes the number of male shopkeepers with shop located in the outskirts.

The first term after the equality sign can be written as:

N_S(male, outskirts, weak) = N_S(outskirts, weak) -
N_S(female, outskirts, weak).

The last term in this equality can be written as:

N_S(female, outskirts, weak) = N_S(female, weak) -
N_S(female, center, weak).

Next we combine the three relations and we obtain:

N_S(male, outskirts) = N_S(outskirts, weak) - N_S(female, weak) +
N_S(female, center, weak) +
N_S(male, outskirts, strong).

Up to now this exposition is independent of the actual cell values. However, when we replace some of the terms in the last relation by the actual values from the Tables 6.22, 6.23 or 6.24 the following happens:

$6 = 9 - 3 + N_S$(female, center, weak) $+ N_S$(male, outskirts, strong).

This relation tells us that the sum of the last two terms is equal to zero. Since both terms must be nonnegative, it follows that both terms themselves are equal to zero. Hence, there are no male shopkeepers in the outskirts of the town with a strong financial position, or, in other words, all male shopkeepers in the outskirts have a weak financial position! In the same manner we conclude that the financial position of all three female shopkeepers in the center is strong.

The three two-dimensional Tables 6.22, 6.23 and 6.24 have been derived from a three-dimensional table which contained the number of self-employed shopkeepers by both financial position and location of business and sex of the self-employed shopkeepers. We therefore call the Tables 6.22, 6.23 and 6.24 linked tables. In general we will call tables linked if they can be deduced from one higher-dimensional table or base file. In the example above we showed that we could deduce the cell values from the three-dimensional table; as a result sensitive information could be disclosed. If only two of the Tables 6.22, 6.23 and 6.24 were published, then it would not be possible to derive the cell values of the three-dimensional table exactly. Hence, a way of protecting the information from the three-dimensional table could be to publish only two instead of three two-dimensional subtables.

With different cell values in the Tables 6.22, 6.23 and 6.24 it might have been impossible to determine the cell values of the three-dimensional table exactly. At the moment no general rule is available which states when the cell values can be re-calculated exactly. What we can do is to indicate how information from a higher-dimensional table can be derived from a set of linked lower-dimensional tables, which all can be obtained from the original higher-dimensional table. For more details we refer to Section 7.5 and [18].

7
Tabular Data: Backgrounds

You cannot win without sacrifice.
—Charles Buxton

7.1 Introduction

In this chapter some theoretical aspects of the SDC for tabular data are examined. It gives some background information on issues raised in the previous chapter, as well as one or two things that have not been discussed there.

The (n,k)-dominance rule and the (p,q)-prior-posterior rule are both examples of a more general class of sensitivity measures, namely the linear subadditive sensitivity measures. Although we do not discuss this class of measures in detail, we do in Section 7.2 show that both the (n,k)-dominance rule and the (p,q)-prior-posterior rule can be written in terms of a linear function.

To protect a sensitive cell it is necessary to ensure that the corresponding feasibility interval is sufficiently wide. The feasibility interval of such a sensitive cell can be determined by solving two linear (or integer) programming problems. This is demonstrated in Section 7.3. In the same section, we examine a rather simple heuristic to determine the secondary suppressions. This heuristic is, theoretically, not as good as some other methods for secondary suppression, but these other methods are often quite complicated. The heuristic we describe, the hypercube method, is due to Repsilber (*cf.* [60]) and can be applied to tables of an arbitrary dimension (provided these tables are not too large). It has the additional advantage that it is easy to understand. Other methods for secondary suppression are not explained because they are mathematically too difficult to include in this book.

Algorithms for controlled rounding are examined in Section 7.4. We study two such algorithms, namely an algorithm by Fellegi (*cf.* [35]) and an algorithm by Cox (*cf.* [11]). The algorithm by Fellegi can only be applied to one-dimensional tables, whereas the algorithm by Cox can be applied to both one-dimensional and two-dimensional tables. Unfortunately, no simple algorithms for controlled rounding of tables of dimension three and higher are known. In fact, it can be demonstrated that there are some three-dimensional tables for which controlled rounding is impossible.

In Section 7.5 some aspects of linked tables are examined. It turns out that for linked tables linear (or integer) programming problems similar to those that are examined in Section 7.3 have to be solved in order to check whether these linked tables are sufficiently safe.

7.2 Sensitivity Measures

The first step when producing a safe table is to determine which cells are sensitive. This information cannot be published unaltered. Determining the sensitivity of a cell can be done in various ways. In this section we show that the (n,k)-dominance rule and the (p,q)-prior-posterior rule are in fact elements of a larger class of sensitivity measures, namely the class of linear subadditive measures.

First we introduce some notation. Let X be and N_X the number of contributors to that cell. The contribution of the i-th contributor is denoted by x_i, $i = 1, \ldots, N_X$. Each contribution x_i is assumed to be nonnegative. Note that for frequency tables $x_i = 1$ for all i. Tacitly we shall assume that the contributions to X are ordered descendingly, that is $x_1 \geq x_2 \geq \cdots \geq x_{N_X} \geq 0$.

Our aim is to express the sensitivity of a cell X by means of a *sensitivity function* S, which is assumed to depend on the individual contributions x_i of the cell total of X. The cell X is considered *sensitive* if and only if $S(X) > 0$. The function S is called *linear* when it depends on the x_i's in a linear way. Finally, the function S is called *subadditive* when the following relation holds for any two cells X and Y and their union $X \cup Y$:

$$S(X \cup Y) \leq S(X) + S(Y).$$

The reason why subadditivity is considered an important property of a sensitivity function is that one usually considers the union of two non-sensitive cells to be non-sensitive again. Subadditive measures possess this property, because when $S(X) \leq 0$ and $S(Y) \leq 0$, then it follows that $S(X \cup Y) \leq 0$, *i.e.* $X \cup Y$ is non-sensitive if X and Y are non-sensitive.

The (n,k)-dominance rule was introduced in Section 6.2. The parameter n is a nonnegative integer and $0 \leq k \leq 100$. In terms of the sensitivity measure S_{dom}, defined as

$$S_{dom}(X) = \sum_{i=1}^{N_X} x_i - (k/100) \sum_{i=1}^{n} x_i \qquad (7.1)$$

the (n,k)-dominance rule can be stated as: a cell X is sensitive if and only if $S_{dom(X)} > 0$. Note that cells X with $n > N_X$ are sensitive under the (n,k)-dominance rule, for any value of k. The function S_{dom} is linear and subadditive.

For the (p,q)-prior-posterior rule which was also introduced in Section 6.2 a similar function $S_{p,q}$ can be derived. We assume that any respondent can estimate the contribution of another respondent to the value of a cell of a table to within q percent of its respective value before this table has been published. A cell of this table is considered sensitive if after the publication of the table it is possible for some respondent to estimate the contribution to the value of the cell of another respondent to within p percent of its original value. It can be shown that a cell is sensitive according to this definition if and only if the second largest respondent is able to determine the contribution x_1 of the largest respondent to within p percent of its value after publication of the table.

We have assumed that the second largest respondent is able to estimate each of the values $x_1, x_3, \ldots, x_{N_X}$ to within q percent. So, by subtracting his own contribution x_2 from the value $T(X)$ of the cell, and by using his prior knowledge, the second largest contributor obtains that x_1 is at most

$$x_1 + (q/100) \sum_{i=3}^{N_X} x_i \qquad (7.2)$$

and that x_1 is at least

$$x_1 - (q/100) \sum_{i=3}^{N_X} x_i \qquad (7.3)$$

The condition that these estimates should differ more than p percent from x_1 in order for the cell to be non-sensitive implies that

$$(q/100) \sum_{i=3}^{N_X} x_i \geq (p/100)x_1 \qquad (7.4)$$

should be satisfied. A linear subadditive sensitivity measure corresponding to (7.4) is given by

$$S_{p,q}(X) = (p/100)x_1 - (q/100) \sum_{i=3}^{N_X} x_i. \qquad (7.5)$$

As usual a cell X is considered sensitive if and only is $S_{p,q}(X) > 0$.

More on linear subadditive sensitivity measures can be found in [10], where the general properties of this class of sensitivity measures are examined.

7.3   Secondary Cell Suppression

When a table needs to be protected one usually starts by redesigning this table, *i.e.* by collapsing some rows or columns. Because protecting a table by table redesign only, leads to too much information loss in many cases, one generally applies another method in combination with table redesign. This other method should protect the cells that remain sensitive after table redesign.

One of these methods to protect sensitive cells is to suppress the corresponding cell values. In tables with marginal totals this is not necessarily enough, because due to the information available in the margins and the other cell values, it could be possible to re-calculate the suppressed values. A solution to solve this problem is to suppress additional cells, *i.e.* cells that were not considered sensitive. This is called secondary suppression. When in a table cells are suppressed secondarily one has to make sure that, on the one hand, the information loss due to this operation is minimized, whereas on the other hand it should not be possible to re-calculate the suppressed values with too great a precision. The possibility to re-calculate suppressed values with a certain precision can for instance occur in tables with restrictions on the range of possible values. Important examples in practice are tables with nonnegative cell values. In the remainder of this section we shall assume that we only deal with such tables. The information loss due to cell suppressions can be measured in various ways; for instance one can take the number of suppressed cells or the total of the suppressed values (*cf.* Section 6.2).

7.3.1   Computing Feasibility Interval

When some values of cells are suppressed in a table with nonnegative cell values, then it is still possible to compute the ranges in which these suppressed values must lie (*cf.* Section 6.4.3). Such a range for a suppressed cell is called the *feasibility interval* for this cell. The feasibility intervals for the sensitive cells should be sufficiently wide, *i.e.* each feasibility interval should contain a corresponding interval which is prescribed by the SDC rules. This prescribed interval is called the *protection interval* corresponding to the sensitive cell. Instead of prescribing a particular protection interval the SDC rules frequently prescribe only the length of this interval.

A feasibility interval for each cell in a table containing suppressed cell totals can be calculated by solving two Linear Programming problems, of the type

$$\left.\begin{array}{c} \min \\ \max \end{array}\right\} x_i, \quad \text{such that} \quad Ax = b, \ x \geq 0 \tag{7.6}$$

where x is a column vector in which each component x_i corresponds to the unknown value of a suppressed cell i (and each suppressed cell is represented by an x_i), and A and b denote a 0-1-matrix and a vector, respectively, that result from the additivity constraints for the table (*i.e.* row-wise and column-wise the cell values add up to the corresponding marginal values).

Note that in case of a frequency count table two Integer Programming (IP) problems, rather than LP problems, should be solved. These IP problems are given by 7.6 and the additional constraint that each component x_i is integer.

To illustrate the method to determine the feasibility intervals of the sensitive cells consider Table 6.10 on page 96. It is assumed that the value of the cell corresponding to activity II and region C is sensitive. The value of this cell is denoted by x_2. We can deduce the following set of equations from this table:

$$\begin{array}{rcl} x_1 + x_2 & = & 30, \\ x_1 + x_3 & = & 25, \\ x_2 + x_4 & = & 34, \\ x_3 + x_4 & = & 29, \\ x_i & \geq & 0 \quad (i = 1, 2, 3, 4). \end{array} \tag{7.7}$$

Our objective is to find the minimal and maximal values the unknowns x_i, $i = 1, \ldots, 4$, which are defined on page 96, can assume under these constraints. These values can be calculated for the sensitive cell x_2 by solving two LP problems, namely

$$\left.\begin{array}{c} \max \\ \min \end{array}\right\} x_2 \quad \text{under the constraints (7.7).} \tag{7.8}$$

The solution to the first LP problem is given by 30 and the solution to the second problem by 5. In a similar way one can determine the feasibility intervals of the other suppressed cells. The resulting intervals for the suppressed cells in Table 6.10 are given in Table 6.19 on page 102.

For a simple suppression pattern like the one shown in Table 6.10 it is not really necessary to solve a number of LP problems. It is also possible to compute the feasibility intervals in the same way as this can be done in the hypercube method for such a case. The hypercube method is explained later in this section.

The method sketched above to determine the feasibility intervals of the suppressed cells can be applied only after cell suppressions are proposed. A way to propose a set of cell suppressions is sketched below.

Suppose a two-dimensional $m \times n$ table is given. Define dummy variables x_{ij} by

$$x_{ij} = \begin{cases} 0 & \text{if cell } (i,j) \text{ is not suppressed} \\ 1 & \text{if cell } (i,j) \text{ is suppressed.} \end{cases} \tag{7.9}$$

It is clear that exactly one suppression in a row or column may not occur, because the value of that cell can be re-computed easily. Hence, necessary, but not necessarily sufficient, conditions for the dummy variables x_{ij} are given by

$$\sum_{i=1}^{m} x_{ij} \neq 1 \quad \text{for all } j = 1, \ldots, n \tag{7.10}$$

$$\sum_{j=1}^{n} x_{ij} \neq 1 \quad \text{for all } i = 1, \ldots, m. \tag{7.11}$$

In order to minimize the information loss due to suppressions the following target function should be minimized:

$$\min \sum_{i=1}^{m} \sum_{j=1}^{n} w_{ij} x_{ij}, \tag{7.12}$$

where w_{ij} is a weight to measure the information loss due to suppression of cell (i,j) (cf. Section 6.4.2).

When the feasibility intervals turn out to be too narrow other cell suppressions must be suggested. This may be quite laborious, because in many cases it is not clear how the proposed suppressions should adapted. Local search techniques, such as simulated annealing ([72]) may be useful in constructing alternative suppression patterns.

Instead of dividing the cell suppression problem into two steps one can also try to solve the cell suppression problem in only one step. In that case the problem of determining the secondary cell suppressions in tables should be formulated as a mixed integer programming problem of the following type: the goal is to find secondary suppressions in a table with given primary suppressions, that minimize an "information loss" function, under the restriction that each feasibility interval associated with a primary suppressed cell contains the, with the corresponding cell associated, protection interval. This problem is called *the secondary cell suppression problem*. A precise mathematical formulation of the secondary cell suppression problem is not presented here, because this is rather involved.

It can be shown that the secondary cell suppression problem is intractable, in the sense that it is likely that there is no efficient way to solve an arbitrary instance of this problem. In practice one therefore has to settle for less, namely for heuristics that not necessarily result in optimal solutions (in the sense of minimum information loss) but that can be efficiently executed. That the optimum is not necessarily reached may be less dramatic than it sounds, because the "real" information loss is something that is rather difficult to quantify.

The secondary cell suppression problem is extensively discussed in [38] and [43]. Heuristics have been formulated for particular tables, in particular two-dimensional ones (cf. [12] and [31], [43], [44]) that are not always easy to generalize to higher-dimensional tables. Neural networks to solve the problem heuristically have been suggested by Wang et al. (cf. [76]). A heuristic that can be applied to tables of arbitrary dimension is described by Repsilber (cf. [60] and [61]) and [21]. We shall refer to this method as the hypercube method. As it seems to be a promising method, and also one that is relatively easy to explain we shall briefly describe it in the remainder of this section. The aim of the hypercube method is to minimize the information loss due to suppressions while guaranteeing that the feasibility intervals of the sensitive cells are sufficiently wide.

7.3.2 The Hypercube Method

What makes the secondary cell suppression problem so tough is that the number of secondary cell suppression patterns to check can be so big. To illustrate this consider a two-dimensional table of size $n \times n$, which has exactly one primary suppressed cell in each row and in each column. Possibly after rearranging the rows and columns of this table we may assume that the primary suppressed cells are on the main diagonal of the table (viewed as a matrix). Assuming that the margins are given with this table as well, we can conclude that it is necessary to suppress (at least) n secondary cells in this table, such that in each row and each column there will be two suppressed cells. Consider the case that in each row and in each column there has to be exactly one secondarily suppressed cell. The number of possible patterns of secondarily suppressed cells with this property equals

$$n! \sum_{k=0}^{n} \frac{(-1)^k}{k!} \rightarrow n!/e \tag{7.13}$$

which is in fact the number of permutations π of n objects which have no fixed points, i.e. with $\pi(i) \neq i$ for all $i = 1, \dots, n$. (In combinatorial enumeration this is called the problem of derangements.) Equation 7.13 shows that the number of possible patterns can be extremely large, even for moderately large values of n. And this is possibly only a subset of the patterns that have to be checked, when taking requirements about the protection intervals for each primary suppressed cell into account. So it would be necessary to reduce the size of the search space, consisting of all feasible secondary cell suppression patterns. The hypercube method does this by considering the primary suppressed cells consecutively (in some order).

For each such cell a hypercube (a rectangle in case of a two-dimensional table) is chosen, with this cell as one of the corner points. This hypercube is called the suppression hypercube. When a suppression hypercube for a particular sensitive cell is selected its corner points are suppressed. Then a suppression hypercube for another sensitive cell is selected. This process continues until a suppression hypercube has been selected for each sensitive cell.

Such a suppression hypercube is selected by minimizing the loss of information due to suppression. To each cell there is a corresponding weight. This weight is used to measure the loss of information when the cell is suppressed. When the weight of a cell is high, then the loss of information due to suppressing this cell is considered high; when the weight of a cell is low, then the loss of information due to suppressing this cell is considered low (*cf.* Section 6.4.2). After a suppression hypercube has been selected for a particular sensitive cell and its corner points have been suppressed the weights corresponding to these corner points are changed into a large negative value. Because the difference between the weight of a suppressed cell and one that has not been suppressed will be quite large in this case, it is likely that cells that have been suppressed will also be chosen in subsequent suppression hypercubes. As a result hypercubes corresponding to different sensitive cells will share the same cells.

It remains to be checked whether the feasibility intervals of the sensitive cells are sufficiently wide. Basically there are two possibilities: either one selects at each step, *i.e.* for each sensitive cell, the hypercube that leads to the minimum loss of information and checks whether the feasibility intervals are sufficiently wide after all these hypercubes have been selected or one takes into account at each step that the feasibility intervals should be sufficiently wide while selecting a suppression hypercube. In the former case after the suppression hypercubes have been selected one can compute the feasibility interval for each sensitive cell exactly, by solving LP problems of the type 7.6. If each of these feasibility intervals contain the corresponding protection interval, the suppression pattern is feasible; otherwise the pattern is rejected. A problem of this method is that, in case the suppression pattern is rejected because at least one feasibility interval is not wide enough, there is not an easy method to find a suppression pattern for which the corresponding feasibility intervals are sufficiently wide.

The other possibility to select a suppression hypercube corresponding to a sensitive cell is to take into account that the feasibility interval of this cell has to be sufficiently wide. This can be done without solving rather time-consuming LP problems. The idea to avoid these LP problems is in fact the main idea of the heuristic proposed by Repsilber.

Suppose we have chosen an n-dimensional hypercube to prevent the disclosure of the value of a particular sensitive cell. We want to determine whether this hypercube can serve as the suppression hypercube for this sensitive cell. When we step from one corner point to the next along the axes of this hypercube we see that each corner point is either an even or an odd number of steps removed from the sensitive cell. A corner point that is an even number of steps removed from the sensitive cell is called an even corner point, the other corner points are called odd corner points. The sensitive cell itself is an even corner point.

We examine how much we can change the values of the corner points of the n-dimensional hypercube while preserving nonnegativity and additivity. Suppose we subtract a value ε_- from the value of the sensitive cell. This implies that we have to add ε_- to all odd corner points and subtract ε_- from all even corner points in order to preserve additivity of the table. Because the values of the cells have to be nonnegative the maximum value of ε_-, $\varepsilon_{m,-}$, equals the minimum value of the even corner points. Likewise, when we add a value ε_+ to the value of sensitive cell the maximum value of ε_+, $\varepsilon_{m,+}$, is given by the minimum value of the odd corner points.

So, when the corner points of the n-dimensional hypercube are suppressed an attacker can deduce that the value of an even corner point lies between $x_e - \varepsilon_{m,-}$ and $x_e + \varepsilon_{m,+}$, where x_e is the true value of this corner point. The value of an odd corner point lies between $x_o - \varepsilon_{m,+}$ and $x_o + \varepsilon_{m,-}$, where x_o is the true value of this corner point. When the numbers $\varepsilon_{m,-}$ and $\varepsilon_{m,+}$ are large enough the protection offered by the n-dimensional hypercube is considered sufficient. This n-dimensional hypercube is then a candidate to be suppressed and will be called a *candidate suppression hypercube*. If the numbers $\varepsilon_{m,-}$ and $\varepsilon_{m,+}$ are insufficiently large the n-dimensional hypercube is not a candidate to be suppressed. Among the candidate n-dimensional hypercubes the one that leads to a minimum loss of information should be selected. This selected candidate suppression hypercube is then the *suppression hypercube*.

Note that the numbers $\varepsilon_{m,-}$ and $\varepsilon_{m,+}$ are in fact lower bounds for the potential decrease and increase of the sensitive cell value, respectively. The reason for this is that it is assumed that there are only two suppressions in each direction. For instance, when there would be three sensitive cells in one row then the actual potential decrease and increase may be larger than $\varepsilon_{m,-}$ and $\varepsilon_{m,+}$, respectively. For computational simplicity, however, the numbers $\varepsilon_{m,-}$ and $\varepsilon_{m,+}$ are used to determine the feasibility interval of the sensitive cell.

The procedure to find the n-dimensional hypercube that minimizes the loss of information is quite easy: just generate all candidate hypercubes and select the one with the minimum loss of information. Note that the above method to find a suppression pattern will be rather fast, because it avoids solving rather time-consuming LP problems; all that needs to be done is to generate hypercubes and to determine the minimums of the even corner points and of the odd corner points, respectively.

The number of hypercubes that have to generated to protect an n-dimensional table is at most equal to

$$s \times \prod_{i=1}^{n}(m_i - 1),$$

where s equals the number of sensitive cells and m_i the number of categories of variable i. For instance, when a three-dimensional $20 \times 20 \times 20$-table is to be protected containing 5 sensitive cells, then at most 34,295 hypercubes have to be generated. Note that the number of hypercubes that have to be generated tends to grow exponentially in the dimension n of the table.

Instead of using the above method to determine the feasibility intervals one might also consider to solve LP problems of the type 7.6 while selecting a hypercube for a sensitive cell. The advantage of this approach is that already suppressed cells are taken into account while computing the feasibility interval for the sensitive cell for which one is selecting a hypercube. However, the number of LP problems that have to be solved may be too large to handle in practice.

7.4   Stochastic Rounding

The problem of rounding in tables stems from the requirement to preserve the additivity of the original table for the rounded table as well. This requirement generally prevents one from treating each cell value in the rounding process in isolation: simply rounding each element in the table to its nearest "rounded number" (*i.e.* a multiple of some base number), as well as the marginal totals, does not ensure additivity in the resulting table. Consider for instance the following frequency count table:

6	2	8
3	1	4
9	3	12

After rounding each value to the closest multiple of 5 it results in

5	0	10
5	0	5
10	5	10

This shortcoming of conventional rounding — the result simply looks awkward — is not the only fault. More importantly, the method does not necessarily yield safe tables! That is, it may be possible to re-calculate the original table provided the attacker knows that a conventional rounding procedure was used. This can only happen when the rounded is not additive. The reader may verify that in the example just presented the original table can be re-calculated from the rounded one.

This defect of the conventional rounding method was the main impetus to look for other rounding methods that guarantee additivity. The idea of the methods that we shall sketch below is that each value (not a multiple of the base value) is rounded to one of the nearest two multiples of this base value, such that the resulting table is additive. The rounding methods are not deterministic but stochastic. Moreover they have the property that the rounding error has zero expectation. This can be achieved if a number, a, that can be written as

$$a = kb + r,$$

for $0 \leq r < b$, k integer and b the integer rounding base, is rounded to a^* which can either be kb or $(k+1)b$, with the following probabilities

$$\begin{aligned} \Pr[a^* = (k+1)b] &= r/b, \\ \Pr[a^* = kb] &= 1 - r/b. \end{aligned} \tag{7.14}$$

As one easily verifies, we indeed have that

$$Ea^* = kb + r = a. \tag{7.15}$$

In fact it is easy to show that only a^* as defined in 7.14, yields the unbiasedness result in 7.15.

The rounding problem for n-dimensional tables can be formulated as a 0-1 integer programming problem. This is not attempted here. Instead we focus our attention on 1- and 2-dimensional tables. For such tables algorithms have been defined by Fellegi and Cox that do not require 0-1 integer programming problems to be solved. Both methods yield unbiased results and preserve additivity of a table. Moreover, if the value of a cell is an integer of the rounding base, then this value is unaltered; if the value of a cell is not an integer multiple of the rounding base, then this value is rounded to one of the nearest two integer multiples of the rounding base. A rounding that satisfies the above criteria is called an *unbiased zero-restricted controlled rounding* in the literature.

7.4.1 The Method of Fellegi for One-dimensional Tables

Fellegi (*cf.* [35]) discusses a random rounding procedure for a one-dimensional table which is both unbiased and preserves additivity. Let

$$a_1 \quad a_2 \quad \cdots \quad a_n \mid t$$

be a one-dimensional table with nonnegative integers a_i, with $t = a_1 + \cdots + a_n$ the table sum and with

$$a_i = k_i b + r_i, \tag{7.16}$$

for some suitable integers k_i and r_i, $0 \le r_i < b$. Let R be a randomly chosen element from the set $\{1, 2, \ldots, b\}$, and let $S_i = r_1 + \cdots + r_i$ be defined as the cumulated sum of the remainder terms of the first i cells. By definition $S_0 = 0$. The rounding procedure is now defined as follows.

$$a_i^* = \begin{cases} (k_i + 1)b & \text{if } \begin{cases} S_{i-1} < R + \ell b \text{ and} \\ S_i \ge R + \ell b \text{ for some } \ell = 0, 1, 2, \ldots \end{cases} \\ k_i b, & \text{otherwise} \end{cases} \tag{7.17}$$

The rounded table sum, t^*, is calculated by summing the rounded cell values, $i.e.$ $t^* = a_1^* + \ldots + a_n^*$. So, additivity is automatically assured. It can be shown that each a_i^* is a random variable with a distribution as a^* in 7.14, and that t^* is one of the two nearest multiples of b of t.

An illustration of Fellegi's rounding procedure can be found in Table 7.1. The numbers a_i have to be rounded to multiples of 5. For all possible choices of R in this case the resulting rounded tables are presented. Note that knowledge of R, the only stochastic element in Fellegi's procedure, completely determines the resulting rounded table.

TABLE 7.1. Fellegi's procedure. Rounding base is 5

			R=1	R=2	R=3	R=4	R=5
a_i	r_i	S_i	a_i^*	a_i^*	a_i^*	a_i^*	a_i^*
12	2	2	15	15	10	10	10
5	0	2	5	5	5	5	5
3	3	5	0	0	5	5	5
18	3	8	20	20	20	15	15
23	3	11	25	20	20	25	25
Total 61			65	60	60	60	60

The rounded values of $a_1 = 12$ for all five possible choices of R are presented in the first row of Table 7.1. Note that the value 10 appears three times in this row, and 15 two times. The expectation of the rounded value of a_1 therefore equals 12. A similar result holds for the other values in the table. This illustrates that the method of Fellegi is an unbiased one.

7.4.2   The Method of Cox for Two-dimensional Tables

Cox ($cf.$ [11]) describes a method for controlled random rounding in two-dimensional tables. To understand this method we must introduce the concept of a cycle in a two-dimensional table.

A *cycle* in a two-dimensional $m \times n$-table A (including marginal totals) is by definition a sequence of distinct pairs of indices (i_1,j_1), (i_2,j_2),..., (i_t,j_t) satisfying

$$1 \leq i_s \leq m, 1 \leq j_s \leq n \quad \text{for} \quad s = 1,...,t, \tag{7.18}$$

$$\begin{array}{lll} j_{s+1} = j_s & \text{and} \quad i_{s+1} \neq i_s & \text{if} \quad s \quad \text{is even} \\ i_{s+1} = i_s & \text{and} \quad j_{s+1} \neq j_s & \text{if} \quad s \quad \text{is odd.} \end{array} \tag{7.19}$$

and

$$j_t = j_1, \tag{7.20}$$

or satisfying 7.18,

$$\begin{array}{lll} j_{s+1} = j_s & \text{and} \quad i_{s+1} \neq i_s & \text{if} \quad s \quad \text{is odd} \\ i_{s+1} = i_s & \text{and} \quad j_{s+1} \neq j_s & \text{if} \quad s \quad \text{is even.} \end{array} \tag{7.21}$$

and

$$i_t = i_1. \tag{7.22}$$

The *distance* between two entries, say (i_x, j_x) and (i_y, j_y), on a cycle is the number of steps, $|y - x|$, that have to be made on the cycle to go from one entry to the other.

A *cycle of unrounded entries* in a two-dimensional table is by definition a cycle of which no entry is equal to an integer multiple of the rounding base.

Now that we have defined cycles in a two-dimensional table, we are able to explain Cox' method. Suppose that the internal cell values of a two-dimensional table A are given by a_{ij} ($i = 1,..,m-1; j = 1,..,n-1$), where a_{ij} denotes the value in row i and column j of the table A. The marginal row total of the i-th row is given by $\sum_q a_{ijq}$, the marginal column total of the j-th column is given by $\sum_p a_{pj}$. The grand total is given by $\sum_p \sum_q a_{pq}$. As the rounding base we use b. Suppose that x is a number that is not an integer multiple of b. In that case we denote the smallest integer multiple of b larger than x by x_+ and the largest integer multiple of b smaller than x by x_-. In other words, when x is rounded up the resulting rounded number is x_+, when x is rounded down the resulting rounded number is x_-.

The first step in the method of Cox is to replace the $m \times n$-table A (including marginal totals) by an $(m+1) \times (n+1)$-table A' with internal cell values a'_{ij} given by

$$a'_{ij} = \begin{cases} a_{ij} & \text{for} \quad 1 \leq i \leq m \quad \text{and} \quad 1 \leq j \leq n \\ -\sum_p a_{pj} & \text{for} \quad i = m+1 \quad \text{and} \quad 1 \leq j \leq n \\ -\sum_q a_{iq} & \text{for} \quad 1 \leq i \leq m \quad \text{and} \quad j = n+1 \\ \sum_p \sum_q a_{pq} & \text{for} \quad i = m+1 \quad \text{and} \quad j = n+1. \end{cases} \tag{7.23}$$

The marginal row totals, the marginal column totals and the grand total of table A' are all equal to zero. It is clear that, when we succeed in finding an unbiased zero-restricted controlled rounding for table A', we have also found an unbiased zero-restricted controlled rounding for the original table A.

Now we choose any entry of table A' that is not an integer multiple of the rounding base, say a'_{kl}. This entry is called the *basic entry*. We also select a cycle of unrounded entries containing the basic entry a'_{kl}. Note that such a cycle can always be selected.

The entries on the selected cycle can be subdivided into two subsets, namely the subset consisting of those entries a'_{ij} for which the distance on the cycle with a'_{kl} is even and the subset consisting of those entries for which the distance on the cycle with a'_{kl} is odd. The elements of the former subset will be called *the even entries on the cycle* and the elements of the latter subset will be called *the odd entries on the cycle*. Note that, because each cycle in a two-dimensional table consists of an even number of entries, a'_{kl} itself is an even entry.

Now note that, when we add a number ε to the even entries on the cycle, then we have to subtract the same number ε from the odd entries on the cycle in order to ensure that the rows and columns of table A' sum up to zero. Likewise, when we add a number ε to the odd entries on the cycle, then we have to subtract ε from the even entries on the cycle. The number ε that is added to or subtracted from the even entries is chosen in such a way that it satisfies the two following requirements:

1. At least one entry is rounded to an integer multiple of the rounding base b when ε is added to or subtracted from the even entries.

2. Each other unrounded entry with a value x is transformed into an entry with a value that lies between x_+ and x_-, i.e. the nearest two integer multiples of b.

As far as the even entries on the cycle are concerned we find that the maximum value that may be added to or subtracted from the even entries, $d_{a,e}$ and $d_{s,e}$ respectively, while satisfying the above criteria 1 and 2 are given by

$$d_{s,e} = \min \{x_+ - x| \; x \text{ is an even entry on the cycle}\}.$$

and

$$d_{s,e} = \min \{x - x_-| \; x \text{ is an even entry on the cycle}\}.$$

As far as the odd entries on the cycle are concerned we find that the maximum value that may be added to or subtracted from the odd entries, $d_{a,o}$ and $d_{s,o}$ respectively, while satisfying the above criteria 1 and 2 are given by

$$d_{a,o} = \min \{y_+ - y| \; y \text{ is an odd entry on the cycle}\}$$

and

$$d_{s,o} = \min \{y - y_-| \; y \text{ is an odd entry on the cycle}\}.$$

As the same value that is added to the even entries must be subtracted from the odd entries and vice versa, one may either add the minimum of $d_{a,e}$ and $d_{s,o}$, say D_{add}, to the even entries, or subtract the minimum of $d_{s,e}$ and $d_{a,o}$, say D_{sub}, from the even entries. So, one may either add D_{add} to the even entries and subtract this value from the odd entries, or one may subtract D_{sub} from the even entries and this value to the odd entries. Either one of the actions results in at least one entry being rounded to an integer multiple of b. For example, when D_{add} is added to the even entries, then one either add $d_{a,e}$ to the even entries (so that at least one even entry is rounded up), or one subtracts $d_{s,o}$ from the odd entries (so that at least one odd entry is rounded down). Similarly, for the subtraction of D_{sub} from the even entries.

Because at least one of the unrounded entries is transformed into a rounded one when one of these two actions is executed, the entire table A' may be rounded by repeatedly executing one of the actions. Furthermore it is clear that each unrounded entry is rounded to one of the two nearest integer multiples of the rounding base. As each marginal row total and each marginal column total of a (partly) rounded version of A' remains zero throughout the procedure, the additivity of the original table is preserved. It remains to ensure that the procedure is unbiased. This is done by using a probability distribution to determine whether the value D_{add} is added to the even entries on a cycle or whether the value D_{sub} is subtracted from the even entries on a cycle. In order to make this method unbiased this probability distribution should be as follows:

$$\Pr[D_{add} \text{ added to even cells}] \quad = \quad D_{sub}/(D_{add} + D_{sub}),$$
$$\Pr[D_{sub} \text{ subtracted from even cells}] \quad = \quad D_{add}/(D_{add} + D_{sub}). \tag{7.24}$$

It is a rather trivial exercise to demonstrate that this probability distribution results in an unbiased procedure. Namely, the expected change in value of the even entries is given by

$$D_{add} \times \frac{D_{sub}}{(D_{add} + D_{sub})} - D_{sub} \times \frac{D_{add}}{(D_{add} + D_{sub})} = 0.$$

Likewise the expected change in value of the odd entries also equals zero.

TABLE 7.2. Enlarged table

20	50	10	-80	0
8	19	22	-49	0
17	32	12	-61	0
-45	-101	-44	190	0
0	0	0	0	0

To illustrate the method of Cox we execute a possible first step of this method to the data in Table 6.1 on page 90. Table 7.2 shows the enlarged table A'.

We select the entry corresponding to Region A and Activity II as the basic entry. This entry has value 8. As the cycle of unrounded entries we select the cells with values 8, 19, 32 and 17 in that order. The even entries are the cells with values 8 and 32, the odd entries are the cells with values 19 and 17. For $d_{a,e}$, $d_{s,e}$, $d_{a,o}$ and $d_{s,o}$ we find the following values.

$$
\begin{aligned}
d_{a,e} &= \min\{10 - 8, 35 - 32\} &= 2, \\
d_{s,e} &= \min\{8 - 5, 32 - 30\} &= 2, \\
d_{a,o} &= \min\{20 - 19, 20 - 17\} &= 1, \text{ and} \\
d_{s,o} &= \min\{19 - 15, 17 - 15\} &= 2.
\end{aligned}
\tag{7.25}
$$

This leads to the following values for D_{add} and D_{sub}.

$$
\begin{aligned}
D_{add} &= \min\{2, 2\} &= 2, \\
D_{sub} &= \min\{2, 1\} &= 1.
\end{aligned}
\tag{7.26}
$$

The probabilities that D_{add} is added to the even entries or that D_{sub} is subtracted from the even entries are given by

$$
\begin{aligned}
\Pr[D_{add} \text{ added to even entries}] &= 1/3, \\
\Pr[D_{sub} \text{ subtracted from even entries}] &= 2/3.
\end{aligned}
\tag{7.27}
$$

Suppose that the outcome of this random selection is that D_{sub} should be added to the even entries (and subtracted from the odd ones). Table 7.3 shows the result.

TABLE 7.3. Enlarged table after the first step

20	50	10	-80	0
7	20	22	-49	0
18	31	12	-61	0
-45	-101	-44	190	0
0	0	0	0	0

This process can be continued until all entries in the table have been rounded. Table 7.4 gives a possible final result after all steps of Cox's method have been completed.

TABLE 7.4. Investments by enterprises (after rounding)

	Region A	Region B	Region C	Total
Activity I	20	50	10	80
Activity II	10	20	20	50
Activity III	15	30	15	60
Total	45	100	45	190

We note again that, in order to preserve the additivity of the table, the cell values are rounded to one of the two adjacent multiples of five, not necessarily to the nearest one. As a consequence, both the value corresponding to activity III and region A and the value corresponding to activity III and region C are rounded to 15. The actual values are 17 and 12 respectively. From the table we can only deduce that both values are in between 10 and 20.

The method of Cox to round two-dimensional tables can also be applied to one-dimensional tables, of course. Unfortunately, it cannot be applied to three- (and higher-) dimensional tables. One of the basic problems is that a closed path in a three-dimensional table may have an odd length, $i.e.$ it may consist of an odd number of entries. This implies that the trick of adding a value to the even entries and subtracting it from the odd ones or vice versa cannot be applied. In fact, it can be shown that there are three-dimensional tables for which there is no unbiased zero-restricted controlled rounding. An example of such a three-dimensional table is the following one.

$$\text{Level 1:} \quad \begin{array}{cc} 1 & 0 \\ 0 & 1 \end{array}$$

$$\text{Level 2:} \quad \begin{array}{cc} 0 & 1 \\ 1 & 0 \end{array}$$

The rounding base b for the above $2 \times 2 \times 2$-table is 2. Each of the four zero entries is already an integer multiple of the rounding base. Therefore, they should remain zero after rounding. Two of the remaining four entries should be rounded to 2. However, these 2's may not appear in the same row, column or level because the entries must sum up to the rounded marginal totals. It is easy to see that this is impossible.

7.5   Linked Tables

The focus of most of the research of the problem discussed in this chapter is on single, isolated tables. In practice one often encounters the problem that a set of tables is released, derived from the same data set and which are, through common variables, linked to each other. In such cases it is not sufficient to consider each table separately, and control its safety in isolation from the other tables. Instead these tables have to be treated in combination.

In the present section we do not deal with the problem of linked tables in detail, because the area is rather extensive and proper treatment of the problems encountered quickly become fairly technical. Instead we illustrate the kinds of problems one is bound to encounter in this area by a simple, but typical, example. The reader should note that, although every single table is safe (*i.e.* has no sensitive cells), the entire set of tables allows one to make a disclosure. Consider Tables 7.5, 7.6 and 7.7. These tables are similar to Tables 6.22, 6.23 and 6.24, except that the values in the cells are different.

TABLE 7.5. Location of business and sex of self-employed shopkeepers

	Center	Outskirts	Total
Male	21	12	33
Female	16	19	35
Total	37	31	68

TABLE 7.6. Financial position and sex of self-employed shopkeepers

	Weak	Strong	Total
Male	23	10	33
Female	8	27	35
Total	31	37	68

The cell values from the three-dimensional table, in which the number of shopkeepers is given by financial position, sex and location of business, are indicated by x_{ijk}. The subscripts indicate successively the sex, location of business and financial position. A number of relations can be derived from the tables. For instance, the following relations can be derived from Table 7.5:

TABLE 7.7. Location of business and financial position of self-employed shop-keepers

	Weak	Strong	Total
Center	11	26	37
Outskirts	20	11	31
Total	31	37	68

$$\text{cell } (1,1): \quad x_{111} + x_{112} = 21,$$
$$\text{cell } (1,2): \quad x_{121} + x_{122} = 12,$$
$$\text{cell } (2,1): \quad x_{211} + x_{212} = 16,$$
$$\text{cell } (2,2): \quad x_{221} + x_{222} = 19.$$

For example, the first relation states that

$$N_S(\text{male, center}) =$$
$$N_S(\text{male, center, weak})(x_{111}) +$$
$$N_S(\text{male, center, strong})(x_{112}) = 21,$$

where N_S denotes the number of shopkeepers.

From Tables 7.6 and 7.7 similar relations can be derived. The result is a system of 12 equations with 8 unknowns. These 12 equations are linear dependent. They are equivalent to a set of seven linear independent equations in eight unknowns. By solving a number of Integer Programming problems of the following kinds:

$$\left.\begin{matrix}\max\\\min\end{matrix}\right\} x_{ijk},$$

where the x_{ijk}'s have to satisfy the seven linear independent equations, $x_{ijk} \geq 0$, and each x_{ijk} is integer, one can determine in which interval the x_{ijk}'s must lie. In this case, just as in the example of Section 6.6, the values x_{ijk} can be determined unambiguously. The x_{ijk} are given by:

$$
\begin{aligned}
x_{111} &= 11, \\
x_{112} &= 10, \\
x_{121} &= 12, \\
x_{122} &= 0, \\
x_{211} &= 0, \\
x_{212} &= 16, \\
x_{221} &= 8, \\
\text{and } \quad x_{222} &= 11.
\end{aligned}
\qquad (7.28)
$$

Generally, however, the values x_{ijk} cannot be determined unambiguously. Instead only intervals in which the values x_{ijk} must lie can be computed.

In discussing linked tables it is useful to make a distinction between two situations. In the first situation the tables to be produced from the base file is a fixed set, in terms of the variables that will span the tables (but not their categorization). In the second situation this set is not fixed. We discuss both situations below.

7.5.1 Fixed Set of Linked Tables

We shall not discuss this matter in its broadest setting, but only for a special case. This case shows a connection with the discussion in Section 5.4 on microdata, in particular with respect to the part on global recoding. Below we only sketch the connection between both areas.

Suppose that a certain fixed set of quantitative tables is to be produced from a certain base file. These sets are the only ones that will be produced from the base set, and they constitute a summary of the data in the base file that will be published. Suppose that this linked set has to be released in such a way that no sensitive cell may appear in any of the tables to be published. Furthermore suppose that the categorization for none of the variables that span the tables is fixed beforehand. Our task is to produce a safe set of tables by tabular redesign only, and with as little information loss as possible. This problem can be solved in a way that is very similar to producing a safe microdata set by only using global recodes. There are two differences, namely the sensitivity criteria and the measures for information loss.

In case of microdata we would have the requirement that each combination to be checked should occur frequently enough, i.e. more frequently than a given threshold value. In case of tabular data we would use a dominance rule, say.

Likewise in case of microdata information loss would be calculated using a quantity such as entropy (cf. Section 5.4), whereas in case of tables one could use the weights specified for each cell to indicate the "importance" for a weight to be published or suppressed. For instance, if an (n, k)-dominance rule is used it would be plausible to use the ratio of the top k contributions of a cell and its total as such an information loss measure. But other such measures could also be used. This similarity between microdata and linked tables with magnitude data should not lead the readers astray: the fact that a set of linked tables does not have any sensitive cells anymore does not imply that cleverly combining these tables will not yield a disclosure. So, the linked set of tables produced here in similarity to a microdata set is only "safe" in the sense that is does not contain any cells that are sensitive according to a particular dominance rule.

In Table 7.8 an overview of microdata versus linked tables is presented.

TABLE 7.8. Microdata versus linked tables with magnitude data

	Microdata	Linked tables (magnitude data)
Sensitive 'entities'	'rare' combinations	cells sensitive according to a criterion such as a dominance rule
Possible SDC measure	global recoding	table redesign
Information loss measured	entropy	cell weights

7.5.2 Non-fixed Set of Linked Tables

Releasing a non-fixed set of tables from the same base file may create problems in controlling their safety. The solution for the fixed set case cannot be used, because it is not possible to modify previously released tables. In particular one faces administrative difficulties when the solution of the problem is sought at the "table level" and not at the level of the base file. For one has to keep track of each table that has been released so far, and possibly also which user has received which tables. The problems one encounters when trying to ensure the safety of individual records when releasing a multitude of tables from the same source file, or equivalently, allowing users to have access to the data in the sense that they can specify their own tables, is closely related to the problems studied in the area of statistical databases (*cf.* Section 10.5 in [69]).

A better way to ensure the safety of a set of tables that is not fixed in advance would be by making sure that the base file itself is safe. But in case there are no restrictions on the tables that one is allowed to produce from the base file, this would imply that the base file itself, viewed as a table, should be safe. This might only be the case if the base file itself has very limited information content. Another possibility, in theory at least, is that the base file itself is perturbed in a suitable way. A difficulty with this approach is that all kinds of inconsistencies might creep into the data in the base file, that may become manifest in tables released from this perturbed file.

8
Loose Ends

The ripest peach is highest on the tree.
—Roscommon

On the final pages of this book we want to make a few comments on various topics that have not been dealt with, or only briefly mentioned, in the previous chapters. Partly they concern issues that deserve further research into the methodology of SDC, in our opinion. For another part they draw attention to the need for adequate and specialized software for the production of "safe" data, microdata and tabular data alike. This software is evidently of great help for the practitioner, charged with the task of producing "safe" data within a limited amount of time. But such software should also be of interest to the researcher in the area of SDC, allowing him to carry out all kinds of "experiments" on data sets more easily. These, in turn, should have their effects on the theory of SDC.

Software

As a result of the sets of SDC rules for various types of microdata and tabular data that have been formulated at Statistics Netherlands in the past several years (*cf.* Section 3.3), the need was felt for specialized software to assist the SDC practitioners, *i.e.* those charged with the production of "safe" data. In 1992 a start was made with the development of a package that was christened ARGUS (*cf.* [16]). Until recently only beta versions of this package were available. Currently version 1.1 is available as the first stable release. This version does not yet possess the full functionality intended for a general SDC package, but some important features are already available. At this moment version 1.1 is only able to deal with microdata. The extension to tabular data has still to be realized. But even for microdata it has not full functionality. The most important feature is that it allows a user to specify combinations of key variables that have to be generated and checked. For each such combination it can check whether the observed combinations of scores in a microdata file occur frequently enough, relative to user specified threshold values. To eliminate rare combinations that have been identified the user has two options available: interactive global recoding and automatic, optimal local suppression. After execution of these actions a report file will be produced containing a list of modifications of the original microdata file.

Further development of ARGUS (*cf.* [25], [22], [58] and [71]) will extend the functionality of the package in such a way as to be able to deal with sampling weights, the scattering of geographical characteristics, checks on household characteristics in view of possible re-grouping of records belonging to the same households. It will also be investigated to automate the global recoding process, possibly in combination with local recoding. Optimal (automatic) global recoding, however, is considerably more complex than optimal (automatic) local suppressing. Not only is it more complex algorithmically, it also requires more preparation at the metadata level. This concerns in particular the definition of suitable proximity structures on sets of categories of key variables.

In order to increase the generality it is also necessary to incorporate at least one perturbative technique into ARGUS. Several such techniques have already been described in the literature (*cf.* [1], [2], [15], [29], [59]). It is possible to use a similar kind of optimization framework as sketched in Section 5.4: instead of determining an optimal mix of global recodes and local suppressions we should determine the optimal mix of global recodes and (local) perturbations. "Optimal" is in both cases referring to minimum information loss. One can even consider the determination of the optimal mix of all three operations.

For SDC of tabular data some software has been produced at various statistical institutes, but none of this is designed for general use, *i.e.* also outside the premises of the institute where the software has been designed. Often this software is only able to deal with 1- or 2-dimensional tables. In practice, however, there is also a need for higher-dimensional tables as well as for linked tables.

The benefits of general SDC software is not only helpful in producing individual "safe" data sets. For instance, it dismisses the practitioner from the obligation to have particular knowledge of SDC methodology, which he cannot be supposed to possess in many cases. But more than this: it even does not require from such a person to have detailed knowledge of the SDC rules that are in use at a statistical office at a particular time, not to mention how to apply them properly. Changes in the SDC rules can be realized by SDC experts (authorized to do so by responsible management) through modification of certain parameter values, that are inaccessible to the ordinary users. It has more far-reaching potentialities. Within a statistical office it can become a key instrument in coordinating and monitoring the production of safe data sets, through the application of uniform SDC rules. On an even larger scale it can be instrumental in harmonizing SDC practices at various national statistical institutes. So the consequences of such a package can be far-reaching.

The following topics are mentioned in the text and deserve further research:

- *Perturbative Techniques. Cf.* Section 1.3 on page 4. Techniques in which "noise" is added to values in records are clearly interesting and important in practice [2]. Such techniques could be used instead of local suppression. The advantage for the user is that no extra "holes" (missings) are created in the microdata set. The challenge for the designer of such techniques is to make sure that they preserve the integrity of the data. In order to identify the "imputed" values in the data set, they should be flagged. For proper statistical analyses it may be necessary to be able to distinguish observed from imputed values.

  One could even fancy much bolder perturbation techniques which distort entire microdata sets, such that "macroscopically" the results obtained from the distorted and the original file are (almost) the same, although "microscopically" the files differ completely (*cf.* [54], for instance, which employs fuzzy sets to distort a microdata set.) Achieving this goal would be very interesting, although analyzing such files could require the use of special statistical software.

- *Re-identification Risk per Record. Cf.* Section 2.5 on page 21. In order to obtain a basis for the SDC of microdata it is necessary to develop models for re-identification risks of individual records in a microdata file. On the basis of such models, applied to a particular data set, it is possible to develop well-motivated actions to increase the safety of the data set, while retaining as much information as possible.

  Once such a re-identification-risk-per record model is at ones disposal one can apply a similar approach to producing a safe microdata set as taken in Section 5.4. However, in this case the aim is not to eliminate rare combinations by actions such as global recodings or local suppressions, but to reduce the re-identification risk of all records such that the resulting file has only records with a controlled maximum risk for each record. As in the case considered in Section 5.4 this goal should be achieved with as little information loss as possible. In fact, information loss need not be measured in terms of entropy in this case. One could instead use the re-identification risk per record: the aim could be to modify the data in such a way that the sum of the re-identification risk (over all records) should be maximal, while trying to bound the re-identification risk of each record below a certain value α.

- *Sampling Weights. Cf.* Section 4.2.7 on page 58. It should be investigated how much "noise" should be added to the sampling weights. The weights should be disturbed by adding as little "noise" as possible, but enough to ensure that the perturbed weights cannot be used for extraction of additional identifying information.

- *Measuring Information Loss. Cf.* Section 5.4 on page 77 and Section 6.4 on page 94. The usage of entropy as a measure for information loss for microdata seems to be an appropriate one. It should be investigated, however, what probability models to use, and in particular what additional information available in the data set. For tabular data one usually defines weights for the cells from which the cells to be secondarily suppressed are selected. These weights should express the desirability to suppress such a cell in a secondary suppression effort. Some examples of such weights are discussed in Section 6.4. It should be investigated, however, whether there are other ways to formally express the desirability for cells to be suppressed, possibly also using entropy. Such sensitivity measures should be useful for table redesign, that also might be carried out in combination with cell suppression.

- *Proximity Structures on Domains of Key Variables. Cf.* Section 5.4 on page 77. In order to be able to carry out automatic global recodes, it is necessary that such proximity structures are defined for all key variables. Sometimes a suitable proximity structure suggests itself, such as for ordinal variables (*e.g.* age, income class), or for variables with a hierarchical code structure (*e.g.* consumer goods, type of business, profession/occupation, type of education), or geographical variables (*e.g.* regions, assuming that the proximity structure can be derived from the contiguity of regions). In case a natural proximity structure does not present itself one could use suitable techniques from multivariate statistics. Which techniques and how to apply them in practice for the purpose intended in this paragraph, should be investigated.

- *Data Editing and SDC. Cf.* Section 5.4 on page 77. From a formal point of view, checking whether certain combinations of scores appear frequently enough in a data set can be viewed as a macro-edit check. Rare occurrences of such combinations are then viewed as macro-edit violations. Such combinations should be eliminated by appropriate actions, such as local suppressions, global recodings or perturbation techniques. It should be investigated to what extent the parallel between data editing for categorical data and SDC exists and whether it is useful to produce software that is flexible enough to deal with both types of problems.

- *Rounding in a Table.* *Cf.* Section 6.4.3 on page 101 and Section 7.4 on page 122. Rounding procedures that preserve additivity for rounded tables, that are unbiased, and that round each cell value to one of the two nearest multiples of a chosen base number have been defined for tables of dimension 1 or 2. However, for higher-dimensional tables it is not always possible to find a rounded table satisfying these conditions. Suitable relaxations have to be found that allow rounding for such tables.

- *Linked Tables.* *Cf.* Section 6.6 on page 108 and Section 7.5 on page 130. The SDC treatment of a fixed set of linked tables by using only table redesign, and the links that exist with SDC for microdata have been mentioned. What remains to be investigated, however, is how other SDC techniques such as cell suppression and rounding can be applied to sets of linked tables with magnitude data. Moreover, linked tables in which the coding of some variables, which refer to the same attribute, differ, remain to be investigated.

The following topics came up while this book was being written, and therefore they have not been investigated yet.

- *Fingerprinting.* Suppose that we have a microdata set and that we have identified the key variables of this set. Suppose also that we do not have any SDC rules we can check this file against in order to find out whether it is safe or not. Yet we want to get a feeling for the safety of this file. In particular we would like to spot its most risky records, in terms of re-identification. An idea that could be useful is to determine fingerprints of each record. A *fingerprint* is a minimum combination of key values in a record that is unique for this record, compared to some reference population, for instance the file in which the record will be released. It is intuitively clear that if a record has many, relatively short fingerprints it is more risky than another record which has less fingerprints or longer ones.

  Determining for each record in a microdata set the (number of or the size distribution of) fingerprints may be an enormous task, particularly when the number of key variables is not very small or when the number of records in the file is large. The major problem is the number of combinations that have to be checked: for m key variables about 2^m combinations of these variables have to be checked, which even for moderate values of m becomes astronomically large. For a given combination a very fast tabulation-like procedure is needed (it need not really count the frequencies of score combinations that occur, but only whether a combination occurred once or more than once.

For each unique combination the record in which it appears should be
kept as well). This task of finding fingerprints can be parallellized. In
case the number of key variables is too large to consider all possible
combinations, only those consisting of at most k key variables could
be checked, where k is not too big. In a sense the more interesting
fingerprints are the shorter ones, for if we take enough key variables
everybody has a unique combination of scores. An SDC-rule based
on fingerprinting could be the following one: a record is considered
unsafe, and hence may not be released unmodified, if it is unique on
more than n k-dimensional keys.

Another interesting question that should be investigated is how the
number and size distribution of fingerprints for a sample relates to
that of the population from which the sample was drawn.

- *Dominance Rule for Tables with Positive and Negative Individual
  Contributions.* The dominance rule (*cf.* Section 7.2) assumes that all
  individual contributions to a cell total are nonnegative. However in
  practice one may be confronted with tables in which this assumption
  is not fulfilled. An example of this is a table with the profits of companies by region and business type. The question arises how one should
  handle such tables, in order to identify the sensitive cells. Note that
  this is a non-trivial problem, which does not follow immediately from
  the dominance rule, because this rule is not invariant under "translations", *i.e.* the addition of a fixed amount to each contribution to
  a cell total.

- *Estimation of Population Frequencies for Complex Sample Designs.*
  In Section 5.2 it is assumed that the microdata set under consideration has been obtained by simple random sampling. In practice this
  is often not the case either as a result of the design used to draw individuals into the sample or as a result of unit and item nonresponse. It
  should be investigated in what sense such situations warrant a more
  sophisticated population frequency estimation procedure.

- *Complex Microdata.* In considering local suppression and global recoding in a microdata set, one does not automatically take into account that there may exist logical or statistical dependencies between
  variables and records. The existence of such dependencies may signifi-
  cantly reduce the effectiveness of SDC actions, when these dependencies are not explicitly taken into account. In extreme cases the effect
  of a local suppression may be nil, because the suppressed result can
  be deduced from non-suppressed information in the same record. A
  similar point is made in a joke by Heinrich Heine, in Chapter XII of
  his "Ideen – Das Buch Le Grand" in his "Reisebilder" (1826), where
  he puts (in translation):

The German censors — — — — — — — — —
— — — — — — — — — — — — — — — —
— — — — — — — — — — — — — — — —
— — — — — — — — — — — — — — — —
— — — — — — — — — — — — — — — —
— — — — — — — — — — — — — — — —
— — — — — — — — fools — — — — — —
— — — — — — — — — — — — — — — —
— — — — — — — — — — — — — — — —
— — — — — — — — — — — — — — — —
— — — — — — — — — — — — — — — —

Even though the censorship seemingly did its job by suppressing the major part of the article, we feel that the message is clearly conveyed by the remaining words. (Although, of course, we are in no way able to deduce what words have been suppressed in the article.)

Therefore local suppression, global recoding and local perturbation models should be investigated that explicitly take such dependencies into account. Microdata in which these dependencies are explicitly taken into account have been called complex microdata in [27] in which also a first attempt can be found at such formulations. Note that the secondary cell suppression problem in tabular disclosure is essentially due to numerical relations between cell values due to the presence of marginal values. So here is another point of similarity between (complex) microdata and tabular data (with marginals).

- *Parallel Computing.* There are several opportunities to investigate possible benefits of high performance (or parallel) computing in SDC. We mention a few such applications. The first one concerns tabular data. Suppose that a table is given in which some cells have been primarily suppressed, and a secondary suppression pattern has been generated, with at least two suppressed cells for each row, column, *etc.* in which a suppressed cell appears (*cf.* Section 7.3). Now it has to be checked for each primary suppressed cell whether the feasibility interval is wide enough. This requires for each primary suppressed cell that two LP problems have to be solved *cf.* Section 7.3). All LP problems can be solved independently of the others, and hence the LP problems can be solved in parallel. For the generation of secondary suppression patterns (*e.g.* using the hypercube method) that have to be checked, parallel computing might also be useful.

Also for the SDC of microdata parallel computing can be applied to speed things up. Fingerprinting was already mentioned. The checking whether certain combinations of scores occur frequently enough in the population is another application. In both case the 'search' space can be naturally divided into disjoint pieces, each of which can be dealt with independently of the others. The same holds for the application of local and global recodings in a microdata set.

With this shortlist of research topics we want to finish the text, because...

We ought to consider the end in everything.
—La Fontaine.

References

[1] G. APPEL AND J. HOFFMANN, 1993, Perturbation by Compensation. In: *Proceedings of the International Seminar on Statistical Confidentiality*, Dublin.

[2] G. BARNES, 1995, Local Perturbation. *Report, Department of Statistical Methods, Statistics Netherlands*, Voorburg.

[3] J.G. BETHLEHEM, W.J. KELLER AND J. PANNEKOEK, 1990, Disclosure Control of Microdata. *Journal of the American Statistical Association*, **85**, 38-45.

[4] J.G. BETHLEHEM AND W.J. KELLER, 1987, Linear Weighting of Sample Survey Data. *Journal of Official Statistics*, **3**, 141-153.

[5] U. BLIEN, H. WIRTH AND M. MÜLLER, 1992, Disclosure Risk for Microdata Stemming from Official Statistics. *Statistica Neerlandica*, **46**, 69-82.

[6] B.D. CAUSEY, L.H. COX AND L.R. ERNST, 1985, Applications of Transportation Theory to Statistical Problems, *Journal of the American Statistical Association*, **80**, 903-909.

[7] A. CHAUDHURI, 1994, Small Domain Statistics: A Review. *Statistica Neerlandica*, **48**, 215-236.

[8] C.A.W. CITTEUR AND L.C.R.J. WILLENBORG, 1993, Public Use Microdata Files: Current Practices at National Statistical Bureaus. *Journal of Official Statistics*, **9**, 783-794.

[9] L.H. Cox, 1977, Suppression Methodology in Statistical Disclosure Analysis. *ASA Proceedings of Social Statistics Section*, 750-755.

[10] L.H. Cox, 1981, Linear Sensitivity Measures in Statistical Disclosure Control. *Journal of Statistical Planning and Inference*, **5**, 153-164

[11] L.H. Cox, 1987, A Constructive Procedure for Unbiased Controlled Rounding. *Journal of the American Statistical Association*, **82**, 520-524.

[12] L.H. Cox, 1993, Solving Confidentiality Problems in Tabulations Using Network Optimization: A Network Model for Cell Suppression in the U.S. Economic Censuses. In: *Proceedings of the International Seminar on Statistical Confidentiality*, Dublin.

[13] F. Crescenzi, 1993, On Estimating Population Uniques; Methodological Proposals and Applications on Italian Census Data. In: *Proceedings on the International Seminar on Statistical Confidentiality*, Dublin.

[14] T. Dalenius, 1982, The Notion of Quasi-identifiers. *Report PRT 5/1 of the Research Project* Access to Information through Censuses and Surveys.

[15] T. Dalenius and S.P. Reiss, 1982, Data-swapping: A Technique for Disclosure Control. *Journal of Statistical Planning and Inference*, **6**, 73-85.

[16] W.A.M. De Jong, 1993, ARGUS: An Integrated System for Data Protection. In: *Proceedings of the International Seminar on Statistical Confidentiality*, Dublin.

[17] J. De Ree and M. Van Huis, 1993, Synthetic Estimation Based on Finite Population Models. In: *Proceedings on Small Area Statistics and Survey Designs*, Warsaw.

[18] R.E. De Vries, 1993, Disclosure Control of Tabular Data Using Subtables. *Report, Department of Statistical Methods, Statistics Netherlands*, Voorburg.

[19] R.E. De Vries, A.G. De Waal and L.C.R.J. Willenborg, 1994, Distinguishing Rare from Common Characteristics in a Microdata Set. *Report, Department of Statistical Methods, Statistics Netherlands*, Voorburg.

[20] A.G. De Waal, 1993, The Number of Tables To Be Examined in a Disclosure Control Procedure. *Report, Department of Statistical Methods, Statistics Netherlands*, Voorburg.

[21] A.G. DE WAAL, 1994, The Hypercube Method for Suppression in Tables. *Report, Statistics Netherlands*, Voorburg.

[22] A.G. DE WAAL AND A.J. PIETERS, 1995, ARGUS User's Guide. *Report, Department of Statistical Methods, Statistics Netherlands*, Voorburg.

[23] A.G. DE WAAL AND L.C.R.J. WILLENBORG, 1994a, Minimizing the Number of Local Suppressions in a Microdata Set. *Report, Department of Statistical Methods, Statistics Netherlands*, Voorburg.

[24] A.G. DE WAAL AND L.C.R.J. WILLENBORG, 1994b, Principles of Statistical Disclosure Control. *Report, Department of Statistical Methods, Statistics Netherlands*, Voorburg.

[25] A.G. DE WAAL AND L.C.R.J. WILLENBORG, 1994c, Development of ARGUS: past, present, future. *Report, Department of Statistical Methods, Statistics Netherlands*, Voorburg.

[26] A.G. DE WAAL AND L.C.R.J. WILLENBORG, 1995a, Statistical Disclosure Control and Sampling Weights. *Report, Department of Statistical Methods, Statistics Netherlands*, Voorburg.

[27] A.G. DE WAAL AND L.C.R.J. WILLENBORG, 1995b, Local Suppression in Statistical Disclosure Control and Data Editing. *Report, Department of Statistical Methods, Statistics Netherlands*, Voorburg.

[28] A.G. DE WAAL AND L.C.R.J. WILLENBORG, 1995c, Optimum Global Recoding and Local Suppression. *Report, Department of Statistical Methods, Statistics Netherlands*, Voorburg.

[29] D. DEFAYS AND N. ANWAR, 1994, Micro-aggregation: A Generic Method. Paper Presented at the Second International Seminar on Statistical Confidentiality, Luxemburg.

[30] J.C. DEVILLE AND C.E. SÄRNDAL, 1992, Calibration Estimators in Survey Sampling. *Journal of the American Statistical Association*, **87**, 376-382.

[31] F. DUARTE DE CARVALHO, N.P. DELLAERT AND M. DE SANCHES OSÓRIO, 1994, Statistical Disclosure in Two-dimensional Tables: General Tables. *Journal of the American Statistical Association*, **89**, 1547-1557.

[32] G.T. DUNCAN AND D. LAMBERT, 1986, Disclosure-limited Data Dissemination. *Journal of the American Statistical Association*, **81**, 10-28.

[33] S. ENGEN, 1978, *Stochastic Abundance Models*. Chapman and Hall, London.

[34] J. FAGAN, B. GREENBERG AND B. HEMMING, 1988, Controlled Rounding of Three-dimensional Tables. *Statistical Research Division Report Series, Bureau of the Census*, Washington D.C.

[35] I.P. FELLEGI, 1975, Controlled Random Rounding. *Survey Methodology*, **1**, 123-133.

[36] I.P. FELLEGI AND A.B. SUNTER, 1969, A Theory for Record Linkage. *Journal of the American Statistical Association*, **64**, 1183-1210.

[37] W.A. FULLER, 1993, Masking Procedures for Microdata Disclosure Limitation. *Journal of Official Statistics*, **9**, 383-406.

[38] J. GEURTS, 1992, Heuristics for Cell Suppression in Tables. *Report, Department of Statistical Methods, Statistics Netherlands*, Voorburg.

[39] B. GREENBERG, 1990, Disclosure Avoidance Research at the Census Bureau. 1990 Annual Research Conference, Bureau of the Census, Arlington, Virginia.

[40] B.V. GREENBERG AND L.V. ZAYATZ, 1992, Strategies for Measuring risk in Public Use Microdata Files. *Statistica Neerlandica*, **46**, 33-48.

[41] J. HOOGLAND, 1994, Protecting Microdata Sets against Statistical Disclosure by Means of Compound Poisson Distributions (in Dutch). *Report, Department of Statistical Methods, Statistics Netherlands*, Voorburg.

[42] W.J. KELLER AND J.G. BETHLEHEM, 1992, Disclosure Protection of Microdata: Problems and Solutions. *Statistica Neerlandica*, **46**, 5-19.

[43] J.P. KELLY, 1990, *Confidentiality Protection in Two- and Three-dimensional Tables*. Ph.D. thesis, University of Maryland, College Park, Maryland.

[44] J.P. KELLY, B.L. GOLDEN AND A.A. ASSAD, 1990, Controlled Rounding of Tabular Data. *Operations Research*, **38**, 760-772.

[45] M. KENDALL AND A. STUART, 1979, *The Advanced Theory of Statistics*, **2**, Fourth Edition (Griffin, London).

[46] LIS, 1993, LIS Information Guide. *Working Paper 7*, Luxemburg.

[47] C. MARSH, C.J. SKINNER, S. ARBER, B. PENHALE, S. OPENSHAW, J. HOBCRAFT, D. LIEVESLEY AND N. WALFORD, 1991, The Case for Samples of Anonymised Records from the 1991 Census. *Journal of the Royal Statistical Society (A)*, **154**, 305-340.

[48] C. MARSH, A. DALE AND C.J. SKINNER, 1994, Safe Data versus Safe Settings: Access to Microdata from the British Census. *International Statistical Review*, **62**, 35-53.

[49] R.J. MOKKEN, P. KOOIMAN, J. PANNEKOEK AND L.C.R.J. WIL-
LENBORG, 1992, Disclosure Risks for Microdata. *Statistica Neerlandica*, **46**, 49-67.

[50] W. MÜLLER, U. BLIEN, P. KNOCHE, H. WIRTH ET AL., 1991,
Die Faktische Anonymität von Mikrodaten. Metzler- Poeschel Verlag,
Stuttgart.

[51] A. NAVARRO, L. FLOREZ-BAEZ AND J.H. THOMPSON, 1988, Results
of Data Switching Simulation. *US Department of Commerce Bureau
of the Census Statistical Support Division*; Paper Presented to the
American Statistical Association and Population Statistics Advisory
Committee.

[52] G. PAASS AND U. WAUSCHKUHN, 1985, *Datenzugang, Datenschutz
und Anonymisierung - Analysepotential und Identifizierbarkeit von
anonymisierten Individualdaten*. Gesellschaft für Mathematik und
Datenverarbeitung, Oldenbourg-Verlag, Munich.

[53] G. PAASS, 1988, Disclosure Risk and Disclosure Avoidance for Microdata. *Journal of Business and Economic Studies*, **6**, 487-500.

[54] S.K. PANDA AND A. NAGABHUSHANAM, 1995, Fuzzy Data Distortion,
Computational Statistics and Data Analysis, **19**, 553-562.

[55] J. PANNEKOEK, 1992, Disclosure Control of Extreme Values of Continuous Identifiers (in Dutch). *Internal Note, Department of Statistical
Methods, Statistics Netherlands*, Voorburg.

[56] J. PANNEKOEK, 1995, Statistical Methods for Some Simple Disclosure
Limitation Rules. *Report, Department of Statistical Methods, Statistics Netherlands*, Voorburg.

[57] J. PANNEKOEK AND A.G. DE WAAL, 1994, Synthetic and Combined
Estimators in Statistical Disclosure Control. *Report. Department of
Statistical Methods, Statistics Netherlands*, Voorburg.

[58] A.J. PIETERS AND A.G. DE WAAL, 1995, A Demonstration of AR-
GUS. *Report. Department of Statistical Methods, Statistics Netherlands*, Voorburg.

[59] S.P. REISS, 1984, Practical Data-swapping: The First Steps. *Trans-
actions on Database Systems*, **9**, 20-37.

[60] D. REPSILBER, 1993, Safeguarding Secrecy in Aggregative Data. In:
*Proceedings of the International Seminar on Statistical Confidential-
ity*, Dublin.

[61] D. REPSILBER, 1994, Preservation of Confidentiality in Aggregated Data. Paper Presented at the Second International Seminar on Statistical Confidentiality, Luxemburg.

[62] D. ROBERTSON, 1994, Automated Disclosure Control at Statistics Canada. Paper Presented at the Second International Seminar on Statistical Confidentiality, Luxemburg.

[63] G. SANDE, 1984, Automated Cell Suppression to Preserve Confidentiality of Business Statistics. *Statistical Journal of the United Nations, ECE 2*, p. 33-41.

[64] C.J. SKINNER, C. MARSH, S. OPENSHAW AND C. WYMER, 1990, Disclosure Avoidance for Census Microdata in Great Britain, In: *Proceedings of the 1990 Annual Research Conference, Bureau of the Census*, Washington, D.C., 131-143.

[65] C.J. SKINNER, C. MARSH, S. OPENSHAW AND C. WYMER, 1992, Disclosure Control for Census Microdata. *mimeo, Department of Social Statistics, University of Southampton*, Southampton.

[66] C.J. SKINNER, 1992, On Identification Disclosure and Prediction Disclosure for Microdata. *Statistica Neerlandica*, 46, 21-32.

[67] C.J. SKINNER AND D.J. HOLMES, 1993, Modelling Population Uniques. In: *Proceedings on the International Seminar on Statistical Confidentiality*, Dublin.

[68] C. SKINNER, C. MARSH, S. OPENSHAW AND C. WYMER, 1994, Disclosure Control for Census Microdata. *Journal of Official Statistics*, 10, 31-51.

[69] J.D. ULLMAN, 1982, *Principles of Database Systems*, 2nd edition. Computer Science Press, Rockville, Md.

[70] U.S. DEPARTMENT OF COMMERCE, 1978, Report on Statistical Disclosure and Disclosure Avoidance Techniques. *Statistical Policy Working Paper*, 2, Washington D.C.

[71] R. VAN GELDEREN, 1995, ARGUS: Statistical Disclosure Control of Survey Data. *Report, Department of Statistical Methods, Statistics Netherlands*, Voorburg.

[72] P.J.M. VAN LAARHOVEN AND E.H.L. AARTS, 1987, *Simulated Annealing: Theory and Applications*. Reidel, Dordrecht.

[73] P. VERBOON, 1994, Some Ideas for a Masking Measure for Statistical Disclosure Control. *Report, Department of Statistical Methods, Statistics Netherlands*, Voorburg.

[74] P. VERBOON AND L.C.R.J. WILLENBORG, 1995, Comparing Two Methods for Recovering Population Uniques in a Sample. *Report, Department of Statistical Methods, Statistics Netherlands*, Voorburg.

[75] A. VERBEEK, 1983, Linear, Subadditive Measures on Finite, Nonnegative, Nonincreasing Sequences. *Report, Department of Statistical Methods, Statistics Netherlands*, Voorburg.

[76] Q. WANG, X. SUN AND B.L. GOLDEN, 1993, Neural Networks as Optimizers: A Success Story. *Working paper, University of Maryland*, College Park.

[77] L.C.R.J. WILLENBORG, 1988, *Computational Aspects of Survey Data Processing*. CWI Tract 54, Centre for Mathematics and Computer Science, Amsterdam.

[78] L.C.R.J. WILLENBORG, 1993, Discussion Statistical Disclosure Limitation. *Journal of Official Statistics*, **9**, 469-474.

[79] L.C.R.J. WILLENBORG, R.J. MOKKEN AND J. PANNEKOEK, 1990, Microdata and Disclosure Risks, *Proceedings of the 1990 Annual Research Conference, Bureau of the Census, U.S. Department of Commerce*, Washington D.C., 167-180.

Index

adding noise, 6, 8, 16, 17, 22, 24,
    43, 59, 70, 81, 82, 137,
    138
ARGUS, vi, 135, 136

circle of acquaintances, 17, 18, 82
classification of variables, 49, 51
coding
    bottom, 67, 70, 74, 77
    top, 42, 67, 70, 74, 77
collapsing, 5, 11, 24, 88
combination
    common, 23, 50, 52, 54
    minimum unsafe, 70, 78, 79
    rare, 19, 23, 25, 50, 52–54,
        63, 73, 74, 77–79, 133,
        136, 137
    unsafe, 25, 50, 63, 70, 78, 79
Cox's method, 114, 123–125, 128,
    129
cycle, 125–128

data
    dissemination, iii, 29, 30, 32,
        36, 44, 49

editing, 80, 138
    swapping, 6, 17
disclosure
    attribute, 15
    control measure, 9, 27, 50, 63,
        65, 70, 81, 88, 90, 91, 94,
        103
    group, 92
    prediction, 15
    re-identification, v, 2, 14, 15,
        17, 20, 54, 64
    risk, 1, 13, 16–18, 20, 21, 23,
        30, 31, 44, 60, 69, 70, 83,
        85, 88, 107
distance, 19, 125, 126

elementary region, 57, 58
estimator
    compromise, 70–73
    direct, 70–73
    interval, 61, 70, 71, 73, 75
    synthetic, 70–73

Fellegi's method, 113, 123, 124
file

150

Lecture Notes in Statistics

For information about Volumes 1 to 24
please contact Springer-Verlag